Power Supplies for LED Driving

Power Supplies for LED Driving

Second Edition

Steve Winder

ELSEVIER

Newnes
An imprint of Elsevier
elsevier.com

Newnes is an imprint of Elsevier
The Boulevard, Langford Lane, Kidlington, Oxford OX5 1GB, United Kingdom
50 Hampshire Street, 5th Floor, Cambridge, MA 02139, United States

Notices
Knowledge and best practice in this field are constantly changing. As new research and experience broaden our
understanding, changes in research methods, professional practices, or medical treatment may become necessary.

Practitioners and researchers must always rely on their own experience and knowledge in evaluating and using any
information, methods, compounds, or experiments described herein. In using such information or methods they
should be mindful of their own safety and the safety of others, including parties for whom they have a professional
responsibility.

To the fullest extent of the law, neither the Publisher nor the authors, contributors, or editors, assume any liability
for any injury and/or damage to persons or property as a matter of products liability, negligence or otherwise, or
from any use or operation of any methods, products, instructions, or ideas contained in the material herein.

Library of Congress Cataloging-in-Publication Data
A catalog record for this book is available from the Library of Congress

British Library Cataloguing-in-Publication Data
A catalogue record for this book is available from the British Library

ISBN: 978-0-08-100925-3

For information on all Newnes publications
visit our website at https://www.elsevier.com/

  Working together
to grow libraries in
developing countries

www.elsevier.com • www.bookaid.org

Publisher: Joe Hayton
Acquisition Editor: Tim Pitts
Editorial Project Manager: Naomi Robertson
Production Project Manager: Nicky Carter
Designer: Mark Rogers

Typeset by Thomson Digital

Contents

Biography

Steve Winder is now a European Field Applications Engineer for Intersil Inc. Steve works alongside design engineers throughout Europe to design circuits using components made by Intersil Inc., a US-based manufacturer of CMOS ICs used for power supply controllers and for analog signal processing.

Prior to joining Intersil Inc., Steve worked for US-based Supertex Inc. in 2002, where he was instrumental in encouraging Supertex's management to start developing LED drivers. One of Steve's German customers had started using a relay driver for LEDs and once Steve had explained the technical detail of this application to Supertex's management, they decided to start an applications team to develop LED-specific products. Supertex then invested heavily to become a leader in this field. Microchip acquired Supertex in 2014.

Until 2002, Steve was, for many years, a team leader at British Telecom Research Laboratories, based at Martlesham Heath, Ipswich in the United Kingdom. Here he designed analog circuits for wideband transmission systems, mostly high frequency, and designed many active and passive filters.

Steve has studied electronics and related topics since 1973, receiving an Ordinary National Certificate (ONC) in 1975 and Higher National Certificate (HNC) in 1977 with Endorsements in 1978. He studied Mathematics and Physics part time with the Open University for 10 years, receiving a Bachelor of Arts Degree with First Class Honors in 1989. He received a Master's Degree in 1991, in Telecommunications and Information Systems after studying at Essex University. Since 1991, he has continued with self-study of electronics, to keep up-to-date with new innovations and developments.

Preface

Welcome to the second edition of Power Supplies for LED Driving! As in the first edition, the worked examples in this book are based on Microchip (formerly Supertex) integrated circuits (ICs), primarily because of my extensive experience with these. However, in this second edition, I introduce ICs from other suppliers and point out the similarities and differences between them. I have also updated the whole book and added new material, including descriptions of new ICs, new light-emitting diode (LED) driving techniques and chapters on both control systems and LED applications. A few minor errors in the first edition have now been corrected and I apologize in advance, in case I have introduced any new ones.

At the beginning of LED development, only those producing red light were available. But these were quickly followed by more colors: yellow/amber, green, and finally blue light, which triggered an explosion in applications. Applications included traffic lights, vehicle lights, and wall washes (mood lighting). LEDs that produce blue light have been combined with yellow phosphor to create white light, which is now one of the most popular colors for general lighting. The amount of light available from LEDs has also increased steadily, and now power levels, up to 20 W, are available using multiple LED die in a single package.

Driving LEDs require a constant current supply. Driving a single LED, or a long string of LEDs connected in series, has relatively few problems when the current is low (maybe 20 mA). High-current LEDs are tougher to drive, requiring 350 mA, 700 mA, 1 A, or a higher rating. Of course, a simple linear regulator could be used if power dissipation was not an issue, or a simple resistor, if current regulation is not critical.

However, in most applications, an efficient switching regulator is used. A switching regulator is essential if the load voltage is higher than the supply voltage (a series-connected LED string), where one needs to boost the voltage. A switching regulator is also needed if the supply voltage has a wide variation, and can be above or below the load voltage at any time, where a boost–buck regulator is needed. But switching means that electromagnetic interference, power dissipation, and parasitic elements have to be considered too. This book describes these in some detail with guidelines and solutions.

This book describes a number of LED driving methods. The main aims of this book are: (1) to show suitable types of LED driver topologies for given applications, (2) to work through some examples, and (3) how to avoid some of the mistakes that some engineers make when creating their own designs. However, the content is not exhaustive and further reading on some peripheral topics will be necessary to obtain a full understanding.

I dedicate this book to Scott Lynch, who died in July 2016 after a long and slow decline in health due to Parkinson's disease. Outside work he was a keen surfer, mostly in Half Moon Bay, but had to give this up when his health declined. Scott was an excellent analog application engineer, being both meticulous and enthusiastic in his work. He was the expert on the switched linear regulator, CL8800, among other things. Scott was a great help to me during the 12 years that I worked with him in Supertex. Scott, may the surf be with you!

Steve Winder
2017

Introduction

Chapter Outline

As a Field Applications Engineer for many years, for one of the pioneering developers of integrated circuits (ICs) for driving power light-emitting diodes (LEDs), I have helped many potential customers in solving design problems. Some have had little or no experience about how to properly drive an LED. Others have had experience with traditional constant voltage power supplies, but LEDs need to be driven with a constant current.

The datasheet of a particular LED will give its current rating, which if exceeded will shorten the expected lifetime. Low power LEDs rated at 20 mA or so can be abused to some extent. However, the power requirements have been increasing; current ratings of 30, 50, 100, 350 mA, and higher are becoming common. If a high power LED is abused, its lifetime will be shortened. Now there are several manufacturers and the power levels are up to 20 W and rising; but these higher powers use LED chip arrays. The names high-bright (HB)-LEDs and ultra-bright (UB)-LEDs are becoming meaningless as the power levels continue to rise. This book will cover all types of LED drivers, from low power to UB-LEDs and beyond.

The unique advantage of LEDs over older styles of lighting is that the color is precise and can be selected for the application. Old filament lamps producing white light are filtered to produce a color, but this is very inefficient because most of the light is blocked by the filter. Obvious applications are traffic lights, using red, amber, and green LEDs. Less obvious applications are lamps for plant growing, where the color affects the type of growth (foliage or fruit). Color control is also used in some alarm clocks, so they wake the user in a controlled manner; color affects mood. A similar application is lighting for seasonally affected disorder (SAD), particularly for people living in the far north or far south, where long periods of darkness during the winter months can lead to depression. Also, LEDs are being used in increasing numbers; in channel lighting (signage), street lights, automotive lighting, mood/atmosphere lighting (color changing "wall wash"), theater lighting for stairs, and emergency exits. More details of these and other applications will be given in Chapter 17.

Is power LED driving simple? No, not usually. In a few cases a linear regulator can be used, which is simple, but most of the cases require a switching power supply with a constant current output. Linear driving is inefficient and generates far too much heat, although for low current applications they can be a good low-cost solution. Fortunately, many IC suppliers provide calculation tools for switching supplies, to help the designer. With a switching supply, the main issues are electromagnetic interference (EMI) and efficiency, and of course cost. The problem is to produce a design that meets legal requirements and is efficient, while costing the least.

1.1 Objectives and General Approach

The approach of this book will be very practical, although some theory is introduced when necessary for understanding of later chapters. It is important to understand the characteristics of components before they can be used effectively. It is also important to understand circuit limitations, to decide whether a particular circuit type is suitable or not to meet the end-equipment's specification. In some cases, costs can be reduced by asking for a change in specification, which can make the circuit simpler. For example, if the power supply voltage is just slightly higher than the LED forward voltage, a linear regulator is cheap and efficient.

In most chapters, I will include a section called "Common Errors." This section will highlight errors that engineers have made, and how these can be avoided, with the hope that readers will not make the same mistakes. It is said that people learn from their mistakes, but it is also true that we can learn from the mistakes of others. Our own mistakes are more memorable, but also more costly!

Usually the first problem for a designer is to choose between different topologies. When is a buck preferred to a buck–boost or a boost? Why is a Cuk boost–buck better than a fly-back type? These choices will be discussed in Chapter 10.

Power supply design equations will be given and example designs of practical supplies will be worked through. With switching power supplies, equations are needed to make the correct component choice; a wrong component can make a poor power supply and require a lot of corrective action. Power LEDs generate a lot of heat in a small area, which makes thermal management difficult, so a colocated power supply should be efficient and will not add too much heating effect.

The implications of changing the calculated component values into standard values, which is more practical, will be discussed. In many cases, customers want to use standard off-the-shelf parts because of ease of purchase and cost. Calculations rarely produce a standard value, so a compromise has to be made. In some cases the difference is negligible. In some cases it may be better to choose a higher (or lower) value. All changes in component value will introduce some "error" in the final result.

Having proven, worked examples in the book will help the reader to understand the design process: the order in which the design progresses. It will also show how the calculated component value compares with the actual value used and will include a description of why the choice was made.

1.2 Description of Contents

In Chapter 2, LED characteristics are described. It is also important to understand the characteristics of LEDs to understand how to drive them properly. One of the characteristics is color; an LED emits a very narrow band of wavelengths so the color is fairly pure. The LED color is determined by the semiconductor materials, which also affect the voltage drop across the LED while it is conducting, so a red LED (a low energy color) has a low forward voltage drop and a blue LED (a high energy color) has a high forward voltage drop. The voltage drop also varies with the current level because there is internal resistance that drops part of the voltage. But the current level determines the light output level: higher current gives higher luminosity from a given LED. The light output from an LED is characterized by both intensity and the angle of beam spreading.

But LED development was driven by the requirements of the end applications in many cases, so Chapter 2 also describes these, but not in great detail. More details of applications will be given at the end of this book, in Chapter 17. The description of some LED applications will show the breadth of the LED driving subject and how LED's physical characteristics can be used as an advantage.

Chapter 3 will show that there are several ways to drive LEDs. As most electronic circuits have traditionally been driven by a voltage source, it is natural for designers to continue this custom when driving an LED. The trouble is that this is not a good match for the LED power requirement. A constant current load needs a constant voltage source, but a constant voltage load (which is what an LED is) needs a constant current supply.

So, if we have a constant voltage supply, we need to have some form of current control in series with the LED. By using a passive series resistor, or active current regulator circuit, we are trying to create a constant current supply. In fact, a short circuit in any part of the circuit could lead to a catastrophic failure, so we may have to provide some protection. Detecting an LED failure is possible using a current monitoring circuit. This could also be used to detect an open circuit. Instead of having a constant voltage supply, followed by a current limiter, it seems sensible to just use a constant current supply! There are some merits of using both constant voltage supply and a current regulator, which will be described in Chapter 4.

Chapter 3 continues, describing features of constant current circuit. If we have a constant current source, we may have to provide some voltage limiting arrangement, just in case the load is disconnected. For example, in a switching boost converter, the output voltage could rise to high levels and damage components in the circuit. Open circuit protection can take many forms. If the circuit failed open, the output voltage would rise up to the level of the open circuit protection limit, which could also be detected.

A short circuit at the load of a linear regulator would make no difference to the current level, so voltage monitoring would be a preferred failure detection mechanism. In some case, the heating effect will be the greatest problem because all of the supply voltage will be across the regulator. However, in a switching buck converter, a short circuit will cause problems because the switching duty cycle is unable to reach zero. This topic is discussed further in Chapter 5.

Another fault detection method, used in switching regulators, is to monitor the switching duty cycle. A short circuit would result in a very short duty cycle and an open circuit would result in a very long duty cycle, so monitoring the duty cycle is an indirect means of fault detection.

Chapter 4 describes linear power supplies, which can be as simple as a voltage regulator configured for constant current. Advantages include no EMI generation, so no filtering is required. The main disadvantage is heat dissipation and the limitation of having to ensure that the load voltage is lower than the supply voltage; this leads to a further disadvantage of only allowing a limited supply voltage range.

Switched linear regulators, where several linear regulators are used in combination, are used in AC mains–powered applications. The regulators are turned on and off as the AC voltage rises and falls. These types of regulators produce low levels of EMI and generally have a good power factor (PF), which are highly desired characteristics. With careful design, good efficiency and good performance are possible. Chapter 4 will discuss the design of such circuits.

Chapter 5 describes the most basic switching LED driver: the buck converter. The buck converter drives an output that has a lower voltage than the input; it is a step-down topology. This type of topology is quite efficient and there are a number of current control methods that are commonly used, such as hysteretic control, synchronous switching, peak current control, and average current control. A number of example driver ICs will be described and compared. The reader will be taken through design processes, followed by example designs.

Chapter 6 describes boost converters. These are used in many applications including LCD backlights for television, computer, and satellite navigation display screens. The boost converter drives an output that has a higher voltage than the input; it is a step-up topology. Battery-powered systems use either inductive boost or charge pump topologies. Higher input voltage systems are usually based on an inductive boost topology. Driver ICs from a number of semiconductor manufacturers will be described. The reader will be taken through the design process, followed by example designs, for both continuous mode and discontinuous mode drivers.

Chapter 7 describes boost–buck converters. These have the ability to drive a load that is either higher or lower voltage compared to the input. However, this type of converter is less efficient than a simple buck or boost converter. These types of topologies are well known as

"Ćuk" or "SEPIC." Again, a number of driver ICs are available for these topologies and they will be described. Design examples of Ćuk and SEPIC will be given.

The boost–buck is popular in automotive applications and battery-operated handheld equipment because the load voltage can be higher or lower than the supply voltage. In automotive applications, the battery voltage varies a lot; low when the starter motor is being operated and high when the alternator is charging the battery. In handheld equipment, the battery voltage starts high, but can fall to low levels when the battery is fully discharged.

Chapter 8 describes nonisolated circuits that have power factor correction (PFC) incorporated into the design. I will start with a typical PFC boost circuit, before moving on to more specialist converters: boost–buck, boost–linear, buck–boost–buck (BBB), and Bi-Bred. These converters are intended for AC input applications, such as traffic lights, streetlights, and general lighting.

The topologies described in Chapter 8 combine PFC with constant current output. In some cases the circuits can be designed without electrolytic capacitors, which are useful for high reliability applications. The efficiency of a circuit with PFC is lower than a standard offline buck converter, but government regulations worldwide require LED lighting to have a good PF if the power level is 25 W or more. This power limit is being reduced and in future could be as low as 5 W. Another reason to have a good PF is cost; utilities charge commercial and industrial customer more if the PF is low.

Chapter 9 describes fly-back converters and isolated PFC circuits. This chapter describes simple switching circuits that can be used for constant voltage or constant current output. A fly-back circuit using two or three windings in the power inductor permits isolation of the output. PFC can be achieved by controlling the input power dynamically over the low frequency AC mains input cycle. But a fly-back using a single winding inductor is actually a nonisolated buck–boost circuit and this is sometimes used for driving LEDs.

Chapter 10 covers topics that are essential when considering a switch mode power supply. The most suitable topology for an application will be discussed. The advantages, disadvantages, and limitations of each type will be analyzed in terms of supply voltage range and the ability to perform PFC. Discussion will include snubber techniques for reducing EMI and improving efficiency and limiting switch-on surges using either inrush current limiters or soft-start techniques.

Chapter 11 describes electronic components for power supplies. The best component is not always an obvious choice. There are so many different types of switching elements: MOSFETs, power bipolar transistors, and diodes, each with characteristics that affect overall power supply performance. Current sensing can be achieved using resistors or transformers, but the type of resistor or transformer is important; similarly with the choice of capacitors and filter components. The performance of operational amplifiers (op-amps), comparators, and high-side current sensors will also be discussed here.

Magnetic components are often a mystery for many electronic engineers and these will be briefly described in Chapter 12. One of the most important physical characteristics from a power supply design point of view, whether designing your own inductors or buying off-the-shelf parts, is magnetization and avoiding magnetic saturation, which will be discussed.

Chapter 12 will also be useful for those designing their own inductors and fly-back transformers. There are different materials: ferrite cores, iron dust cores, and special material cores. Then there are different core shapes and sizes. Some cores need air gaps, but others do not. All these topics will be discussed.

EMI and electromagnetic compatibility (EMC) issues are the subjects of Chapter 13. It is a legally binding requirement in most parts of the world that equipment should meet EMI standards. Good EMI design techniques can reduce the need for filtering and shielding, so it makes sense to carefully consider this to reduce the cost and size of the power supply. Meeting EMC standards is also a legal requirement in many cases. It is of no use having an otherwise excellent circuit that is destroyed by externally produced interference, such as a voltage surge on an AC line or a voltage spike across an automotive DC supply. In many areas, EMC practices are compatible with EMI practices; so fixing one often helps to fix the other.

Chapter 13 also covers MOSFET-driving techniques that, while reducing EMI, will also reduce switching losses. This will increase the efficiency and reliability of the LED driver.

Chapter 14 discusses thermal issues for both the LEDs and the LED driver. The LED driver has issues of efficiency and power loss. The LED itself dissipates most of the energy it receives (voltage drop multiplied by current) as heat: very little energy is radiated as light, although manufacturers are improving products all the time. Handling the heat by using cooling techniques is a largely mechanical process, using a metal heat sink and sometimes airflow to remove the heat energy. Calculating the temperature is important because there are operating temperature limits for all semiconductors.

Another legal requirement is safety, which is covered in Chapter 15. The product must not injure people when it is operating. This is related to the operating voltage and some designers try to keep below safety extra low voltage (SELV) limits for this reason. When the equipment is powered from the AC mains supply, the issues of isolation, circuit breakers, and creepage distance (the space between high and low voltage points on the PCB) must be considered. Some applications, such as swimming pool lighting, have very strict rules for safety (as you would expect).

Chapter 16 covers control systems. These include traditional control like 1–10 V linear dimming and triac-controlled phase-cut dimmers. A description of triac-controlled dimmers will be given to explain why they are so difficult to use with LED lighting. Newer digital control techniques, such as DALI and DMX will be described. Automotive and industrial applications often use the LIN bus or the CAN bus, so some description of these will be given.

Chapter 17 returns to the topic of applications. Applications were briefly described in Chapter 2 to help explain the development of LEDs. In this chapter we describe more applications and in greater detail, while referring to the various LED driver circuits that are suitable. The reasons why some circuit topologies are better than others in a particular application will be discussed.

Characteristics of LEDs

Chapter Outline

Most semiconductors are made by doping silicon with a material that creates a free-negative charge (N-type), or free-positive charge (P-type). The fixed atoms have positive and negative charge, respectively. At the junction of these two materials, the free charges combine and this creates a narrow region devoid of free charge. This "intrinsic region" now has the positive and negative charge of the fixed atoms, which opposes any further free charge combination. In effect, there is an energy barrier created; we have a diode junction.

In order for a P–N junction to conduct, we must make the P-type material more positive than the N-type. This forces more positive charge into the P-type material and more negative charge into the N-type material. Conduction takes place when (in silicon) there is about 0.7-V potential difference across the P–N junction. This potential difference gives electrons enough energy to conduct.

A light-emitting diode (LED) is also made from a P–N junction, but silicon is unsuitable because the energy barrier is too low. The first LEDs were made from gallium arsenide (GaAs) and produced infrared light at about 905 nm. The reason for producing this color is the energy difference between the conduction band and the lowest energy level (valence band) in GaAs. When a voltage is applied across the LED, electrons are given enough energy to jump into the conduction band and current flows. When an electron loses energy and falls back into the low-energy state (the valence band), a photon (light) is often emitted (Fig. 2.1).

2.1 Applications for LEDs

Soon new semiconductor materials were developed and gallium arsenide phosphide (GaAsP) was used to make LEDs. The energy gap in GaAsP material is higher than GaAs, so the light wavelength is shorter. These LEDs produced red color light and were first just used as indicators. The most typical application was to show that equipment was powered, or that

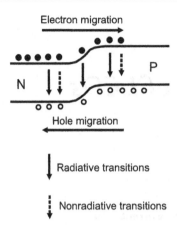

Figure 2.1: Band Diagram of P–N Junction Semiconductors.

some feature such as "stereo" was active in a radio. In fact it was mainly consumer products, such as radios, tape recorders, and music systems that used red LEDs in large numbers.

When yellow and green LEDs became available, the number of applications increased. Now the color could change, to give additional information, or could indicate more urgent alarms. For example, green = OK, yellow = requires attention, and red = faulty. Most important was the ability to have LED lamps in traffic lights.

One characteristic of the light from an LED is that it occupies a narrow spectrum about 20-nm wide; the color is fairly pure. By contrast, a semiconductor laser used for telecommunications occupies a spectrum about 2-nm wide. The very narrow spectrum of a laser is important because, when used with optical fiber systems, the narrow spectral width allows a wide-system bandwidth. In general-purpose LED applications, the spectral width has very little effect.

Another important characteristic of LED light is that current is converted into light (photons). This means that doubling the current doubles the light amplitude. So dimming lights by lowering the current is possible. It should be noted that the specified wavelength emitted by an LED is at a certain current; the wavelength will change a little if the current is higher or lower than the specified current. Dimming by pulse width modulation is a viable alternative used by many people. pulse width modulation dimming uses a signal, typically a frequency range of 100–1000 Hz, to turn the LED on and off. The pulse width is reduced to dim the light, or increased to brighten the light.

The "holy grail" was blue LEDs, which are made from indium gallium nitride (InGaN). When adding colored light, red, green, and blue make light that appears white to the human eye. The reason for only "appearing" white is that the eye has receptors (cones) that detect red, green, and blue. There are big gaps in the color spectrum, but the eye does not notice. White LEDs are usually made using blue LEDs with a yellow phosphor dot over the emitting surface. The yellow phosphor creates a wide spectrum and, when combined with the blue, appears white. Different phosphors produce a range of white from cold white (bluish white) to warm white (yellowish white).

An interesting application for blue LEDs is in dentistry. Illuminating modern resins used in filling materials with blue light can cure them (make them change form, from liquid into solid). The 465-nm wavelength has been found to be close to optimum for this application, although the intensity of the light must be high enough to penetrate through the resin.

Some interesting applications rely on the purity of the LED color. The illumination of fresh food is better with LEDs because they emit no ultraviolet light. Photographic dark rooms can use colors where the film is insensitive; traditionally red-colored incandescent lamps. Even traffic lights must emit a limited range of colors, which are specified in national standards. Incandescent (filament) lamps in these applications were inefficient because they produce white light that is filtered, so most of the light energy is blocked by the filter and thus wasted as heat.

It should be noted that the color of an LED would change as the LED's temperature changes. Some colors change more than others: a customer once told me that the amber-colored LED for a traffic light is very susceptible to color change and it is difficult to keep within specification. The LED temperature can change due to ambient conditions, such as being housed adjacent to hot machinery, or due to internal heating of the LED because of the amount of current flowing through it. The only way to control ambient temperature is to add a cooling fan, or by placing the LED away from the source of heat. Mounting the LED on a good heat sink can help to control internal heating.

In the early days of LED production, all LEDs had a current rating of 20 mA and the forward voltage drop was about 2 V for red, higher for other colors; later low-current LEDs were created that operated from a 2-mA source (for battery-powered applications). One LED manufacturer called Lumileds was created by HP and Philips in 1999 and produced the first 350-mA LED (Luxeon Star). Now there are a number of power LED manufacturers, rated at 350 mA, 700 mA, and 1 A or higher. Power LEDs have been used in increasing numbers; in channel lighting (signage), traffic lights, street lights, automotive, agricultural lighting, mood lighting (color changing "wall wash"), and also in theatres for lighting steps and emergency exits.

However, low-current LEDs in the 20–100 mA range are more efficient: the light output per watt (effectiveness or efficacy) is better for lower-current LEDs. Low-current LEDs are also very much cheaper than high-current LEDs. Since 2010, the trend is to use an array of many low-current LEDs instead of a single high-current LED, to produce more lumens per watt. The power rating of these LED arrays can reach 150 W (e.g., Cree CXA3590 LED). In some

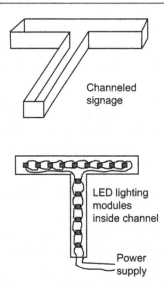

Figure 2.2: Channel Lighting.

lamps this has the benefit of distributing the light over a larger area and thus avoiding the need for lenses or light diffusors. Another advantage is that the distribution of heat is easier with multiple low-current LEDs rather than a few high-power LEDs.

Channel lighting is so called because the LEDs are mounted in a channel (Fig. 2.2). Typically this channel is used to form letters, for illuminated company name signs. In the past, channel lighting used cold-cathode or fluorescent tubes, but these had reliability problems. Health and safety legislation, such as the RoHS Directive, banned some materials, such as mercury that is used in the construction of cold-cathode tubes. So, to cope with the shapes and environmental conditions, the most viable technology is LED lighting.

Traffic lights have used low-power LEDs for some years, but now some manufacturers are using a few high-power LEDs instead. Traffic lights are usually powered from the AC mains supply and good power factor correction is required. Having a high power factor (PF) means that the supply current is in phase with the supply voltage and the harmonic distortion is low. Keeping the PF high helps to reduce the cost of the AC supply because power supply companies charge a premium if the PF is low. Also, some traffic light control systems need a load that has a good PF, for fault detection purposes.

One problem with traffic lights is controlling the wavelength of the yellow (amber) light. Yellow LEDs suffer from a greater wavelength shift than other colors, and this can cause them to operate outside their permitted spectral range. Another problem is making them fail-safe; authorities permit some degree of failure, but usually if more than 20% of the LEDs fail, the entire lamp must be shut down and a fault reported to maintenance teams.

High-ambient temperatures inside the lamp housing can lead to LED driver failures. This is particularly true if the LED driver circuit contains electrolytic capacitors, which vent when hot and eventually lose their capacitance. Some novel LED drivers have been developed that do not need electrolytic capacitors and can operate for several years at high temperatures. Failing LED drivers can give LED lights a bad name; why have LEDs that can work for over 100,000 h if the LED driver fails after a 10,000-h operation?

Streetlights have been built using medium- and high-power LEDs. Although this would seem to be a simple application, high-ambient temperatures and relatively high-power LEDs can give rise to driver problems.

White LEDs are made using a blue LED and a yellow phosphor. The blue photons have high energy and when they combine with the yellow phosphor, the photon energy is reduced. The resultant light has a broad spectrum, with peaks in the blue and yellow regions. One problem is that the high-blue content produces a "cold-white" light. In some cases, white and yellow LEDs are used together to create a "warm-white" light. However, phosphor technology has developed to allow "warm-white" LEDs. In general, warm-white LEDs have a lower efficacy (fewer lumens per watt) because the phosphors used in these LEDs absorb more photons than they emit.

Automotive lighting has many applications; internal lights, head-lights, stop lights, daylight running lights (DRL), rear fog lights, reversing lights, etc. The greatest problem with automotive applications is that the EMI specifications demand extremely low levels of emissions, which are difficult to meet with a switching circuit. The driver circuit is often shielded with a metal box to achieve low-emission levels. Linear drivers are sometimes used to reduce costs, provided that the efficiency is not a critical requirement. Connecting a linear driver to the metal body of the vehicle can be used to dissipate the heat generated.

Automotive stoplights using LEDs have a significant safety advantage over those using filament lamps. The time from current flow to light output in an LED is measured in nanoseconds. In a filament lamp the response time is about 300 ms. At 60 mph (100 km/h), a vehicle travels 1 mile (1.6 km) per minute, or 88 ft./s. In 300 ms, a car will travel over 26 ft. (8 m). Stopping 300 ms sooner, having seen the previous car's brake lights earlier, could save death or injury. Also, LED brake lights are less likely to fail than filament lamps, due to being able to withstand greater levels of vibration.

Agriculture has benefited from using LEDs because the plant growth is affected by color. Green light has little effect on plant growth, but both red and blue light do. So immediately, the efficiency is improved because green light does not need to be produced; we only have to produce wavelengths that affect the plants. One effective wavelength is ~660 nm (red) because this wavelength corresponds to an absorption peak in chlorophyll. So red light helps leaf growth, whereas adding blue light (~450 nm) helps with stem and seed growth (wheat would grow with just red light, but the addition of blue increases the size and quantity of the seeds).

Mood lighting is an effect caused changing the color of a surface and uses human psychology to control people's feelings. It is used in medical facilities to calm patients, and on aircraft to relax (or wake up!) passengers. Blue light is sometimes used by long-distance truck drivers, to help them stay alert. Note that the LCD backlight in some handheld computers produces a significant amount of blue light, which can lead to insomnia if the computer is used at bedtime.

Generally mood lighting systems use red, green, and blue (RGB) LEDs in a "wall wash" projector to create any color in the spectrum. Other applications for these RGB systems include disco lights! One mood lighting application is an alarm clock that has an LED lamp to simulate dawn, with the light level starting very low and gradually increasing. The same manufacturer makes seasonally affected disorder lights to help people affected by depression during the long-winter months (in the Northern hemisphere at least).

Backlighting displays, such as flat screen televisions, also use RGB LED arrays to create a "white" light. In this case the color changes little, ideally not at all. However, a control system is required to carefully control the amount of red, green, and blue to create the exact mix for accurate television reproduction. To create a wide-dynamic range, the intensity of the backlight is dynamically controlled, so for dark scenes the backlight is dimmed and for bright scenes the backlight is brightened. Global dimming means the whole backlight brightness has a single control. Some backlights are divided into sections, for example, six-horizontal bands, so that each section's brightness can be controlled independently. This latter scheme works because most outdoor scenes have a dark ground and a bright sky.

LEDs are also used to backlight computer screens, cashpoint displays, ATM machines, etc., but here the exact color or brightness is less important. In some cases the backlight is used to help create a "corporate" color scheme.

2.2 Light Measure

The total light flux is measured in units of lumens. The lumen is the photometric equivalent of 1 W, weighted to match the normal human eye response. At 555 nm, in the green–yellow part of the spectrum where the eye is most responsive, 1W = 683 lm.

The term candela is also used. This is the light produced by a lamp, radiating in all directions equally, to produce 1 lm/sr. As an equation, 1 cd = 1 lm/sr. A steradian has a projected area of 1 m^2, at a distance of 1 m from the light source. The light from a 1-cd source, at a meter distance, is 1 lux or 1 lm/m^2 (Fig. 2.3).

Light emission efficiency (luminous efficacy) from LEDs is described in terms of lumens per watt. There is some competition between LED manufacturers get the highest luminous efficacy, but when comparing results it is important to make a note of the electrical power

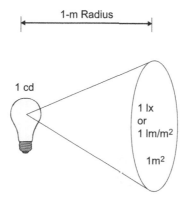

Figure 2.3: Light Measurement.

levels used. It is easier to make an efficient 20-mA LED, than an efficient 700-mA LED. As already mentioned, this is why using LED arrays of multiple low-power LEDs are now preferred over using a few high-power LEDs.

2.3 Equivalent Circuit to a LED

An LED can be described as a constant voltage load. The voltage drop depends on the internal energy barrier required for the photons of light to be emitted, as described earlier, and also on the equivalent series resistance. The energy barrier depends on the color; thus the voltage drop depends on the color. Will every red LED have the same voltage drop? No, because production variations will mean that the wavelength (color) will not be the same, nor will the equivalent series resistance, and thus the voltage drop will have differences. The peak wavelength has typically a ±10% variation.

If there are temperature differences between two LEDs, this will cause differences in their forward voltage drop. As the junction temperature rises, it is easier for electrons to cross the energy barrier. In fact, the voltage drop reduces by approximately 2 mV/°C as the temperature rises. This is a problem if, for example, two LEDs are connected in parallel: the hotter LED will have a lower forward voltage drop, which will mean that is takes a greater share of the total current, which means that it gets even hotter and its forward voltage drop becomes lower, and the current rises further, etc. In other words, thermal runaway. In LED arrays, which have series/parallel connections, good thermal coupling between LED die is essential, and is achieved by mounting the LED die on a ceramic substrate.

LEDs are not really a constant voltage load. As the semiconductor material is not a perfect conductor, some resistance is in series with this constant voltage load (Fig. 2.4). This means that the voltage drop will increase with current. The equivalent series resistance (ESR) of a low-power, 20-mA LED is about 20 Ω, but a 1-W, 350-mA LED has an ESR of about 1 Ω.

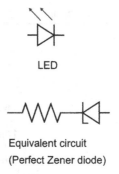

LED

Equivalent circuit
(Perfect Zener diode)

Figure 2.4: Equivalent Circuit for a Light-Emitting Diode (LED).

Manufacturers are continually working to reduce the ESR. The ESR is roughly inversely proportional to the current rating of the LED. The ESR will have production variations too.

The ESR can be calculated by measuring the increase in forward voltage drop divided by the increase in current. For example, if the forward voltage drop increases by from 3.5 to 3.55 V (a 50-mV increase) when the forward current goes from 10 to 20 mA (a 10-mA increase), the ESR will be 50 mV/10 mA = 5 Ω.

In Fig. 2.4, the Zener diode is shown as a perfect device. In reality, Zener diodes also have ESR, which can be higher than the ESR of an LED. For initial testing of an LED driver, a 3.6- or 3.9-V Zener diode (rated for 5-W power) can be used to replace a white LED. If the driver is not working as planned the Zener diode may be destroyed, but this is far less costly than destroying a power LED. As the Zener diode does not emit light, the test engineer will not be dazzled. However, be aware that Zener diodes can get very hot. I built an LED driver test jig using wire-ended Zener diodes mounted in block terminal strips (connected in series for easy connection). The first block terminal strips that I used were made from molded nylon, which melted due to the heat generated by the Zener diodes.

2.4 Voltage Drop Versus Color and Current

The graph in Fig. 2.5 shows how the forward voltage drop depends on the light color and on the LED current. At the point where conduction begins, the forward voltage drop, V_f, is about 2 V for a red LED and about 3 V for a blue LED. The exact voltage drop depends on the manufacturer because of different dopant materials and wavelengths. The voltage drop at a particular current will depend both on initial V_f and on the ESR.

2.5 Common Mistakes

The most common mistake is to base a design on the typical forward voltage drop of the LED, V_f type. This includes connecting strings of LEDs in parallel, with the assumption that the forward voltage drops are equal and the current will share equally between the two or

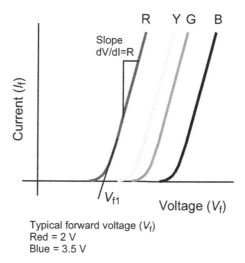

Figure 2.5: Forward Voltage Drop Versus Color and Current.
B, Blue; *G*, green, *R*, red; *V*$_f$, forward voltage drop; *Y*, yellow.

more strings. I have described in this chapter the problem of thermal runaway with parallel connected LEDs, where one LED takes the majority of current because of the lower forward voltage drop when it gets warm. The tolerance on the forward voltage drop is very high. For example, a 1-W white Luxeon Star has a typical V_f = 3.42 V, but the minimum voltage is 2.79 V and the maximum is 3.99 V. This is over a ±15% tolerance on the forward voltage drop!

CHAPTER 3

Driving LEDs

Chapter Outline

3.1 Voltage Source

We have seen in Chapter 2 that a light-emitting diode (LED) behaves like a constant voltage load with low equivalent series resistance (ESR). This behavior is like a Zener diode—in fact Zener diodes make a good test load, rather than using expensive high power LEDs!

Driving a constant voltage load directly from a constant voltage supply is very difficult because it is only the difference between the supply voltage and the load voltage that is dropped across the ESR. But the ESR is a very low value, so the voltage drop will also be low. The LED forward voltage and ESR are also temperature dependent. A slight variation in the supply voltage, or the LED forward voltage, or the ESR, will cause a very large change in current; see curve A in Fig. 3.1.

If the variation in supply voltage and forward knee voltage (V_f) is known, the variation in current can be calculated. Remember that there are variations in LED voltage drop due to both manufacturing tolerances and operating temperature. Most supply voltages from a regulated supply have a 5% tolerance, but from unregulated supplies like automotive power, tolerances are far greater.

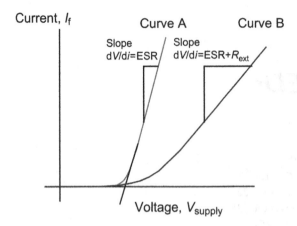

Figure 3.1: LED Current Versus Supply Voltage.

$$I_{min} = \frac{V_{source\_min} - V_{f\_max}}{ESR}$$

$$I_{max} = \frac{V_{source\_max} - V_{f\_min}}{ESR}$$

These equations assume that ESR is constant. In practice, the V_f and voltage drop across ESR are combined since manufacturers quote the voltage drop at a certain forward current. The actual V_f of the LED diode junction can be determined from graphs, or can be measured.

Consider that there is a large difference between the source and load voltage, which is dropped across a high value series resistor. In this case, changes in LED forward voltage cause very little difference between the maximum and minimum LED current. This may be perfectly adequate for low current LEDs, up to 50 mA. However, in high power LED circuits, a large voltage drop across a series resistor will be inefficient and may cause heat dissipation problems. Also, the ESR of LEDs is lower as the power rating increases. As previously described, a standard 20-mA LED may have an ESR of about 20 Ω, but a 350-mA LED will have an ESR of 1 Ω typically. Thus a 1 V difference in supply voltage could increase the LED current by 1 A in a power LED. Even in low current LEDs, the proportional change in current can be very high.

3.1.1 Passive Current Control

Looking again at Fig. 3.1, if the LED voltage drop increases, curve A of the graph shifts to the right. The slope of the graph is unchanged and is just due to the ESR. However, low current LEDs can have a relatively high value resistance added in series, to reduce the slope of the current versus voltage graph; see curve B in Fig. 3.1. Now a change in supply voltage has much less effect on the LED current, I_f.

With a series resistor added we are able to calculate the variation in current, provided that the variation in supply voltage and load voltage is known. In the following equations, the load voltage includes the voltage drop across ESR, at the rated current, so only the external resistor value is needed.

$$I_{min} = \frac{V_{source\_min} - V_{load\_max}}{R_{ext}}$$

$$I_{max} = \frac{V_{source\_max} - V_{load\_min}}{R_{ext}}$$

As an example, let us drive an LED from an automotive supply; this is a nominal 13.5 V, but for this exercise we can set the limits at 12–16 V. Let us select a red LED for tail-lights (Lumileds Superflux HPWA-DDOO), with a forward voltage drop of 2.19–3.03 V at 70 mA forward current. Choosing to connect two LEDs in series, with a series resistor, we have a typical voltage drop of 5 V. So the typical voltage drop at 70 mA needs to be 8.5 V; this means that the series resistor should be 121.43 Ω. The nearest standard value of resistor is 120 Ω, rated at 1 W since we will have a typical power dissipation of 588 mW.

$$I_{min} = \frac{V_{source\_min} - V_{load\_max}}{R_{ext}} = \frac{12 - 6.06}{120} = 49.5 \, mA$$

$$I_{max} = \frac{V_{source\_max} - V_{load\_min}}{R_{ext}} = \frac{16 - 4.38}{120} = 96.83 \, mA$$

At the high limit of source voltage, the LED is overdriven by 38%. But there is almost a 2:1 ratio between I_{max} and I_{min}, so if we increase R by 38% the worst case current levels are 70 mA maximum and only 35.78 mA minimum.

In the previous calculations, the voltage drop across ESR (0.672 V) was included in the minimum and maximum load voltage values, so we ignored ESR. From the manufacturer's data sheet of the Lumileds HPWA-DDOO LED, graphs show that the ESR is about 9.6 Ω. Suppose we now want to operate at a lower current. Using the same example, but operating with a typical LED current of 50 mA, we must modify the results. Now the voltage at the current knee is V_f = 1.518–2.358 V. With a typical 13.5-V supply and 50 mA, the value for V_f is 1.828 V. The total resistance needed is 196.88 Ω, but we already have 9.6 Ω ESR. An external resistor value of 180 Ω is the nearest preferred value for a current of 50 mA.

$$I_{min} = \frac{V_{source\_min} - V_{load\_max}}{ESR + R_{ext}} = \frac{12 - 4.716}{189.6} = 38.42\,\text{mA}$$

$$I_{max} = \frac{V_{source\_max} - V_{load\_min}}{ESR + R_{ext}} = \frac{16 - 3.036}{29.6} = 61.85\,\text{mA}$$

The series resistor has a higher value, so the variation in current is reduced to 1.6:1 ratio. The maximum current is now below the LED current rating of 70 mA.

What happens when we connect LEDs in parallel? Unless the LEDs are matched (or "binned") to ensure the same forward voltage drop, the current through one string could be considerably different from the current through another. Even if the forward voltages are matched initially, differences in power dissipation from one LED to another will cause a temperature difference in the LEDs. As the temperature rises, the forward voltage drop reduces. This causes more current through the LED, the ESR dropping a higher voltage, and so the temperature rises further. At the same time, parallel LEDs will take less current and cool down. We end up with thermal runaway and ultimately LED damage.

When multiple LEDs are used to provide lighting for an application, they are frequently connected in an array, consisting of parallel strings of series connected LEDs. Since the LED strings are in parallel, the voltage source for all strings is the same. However, due to manufacturing variations in forward voltage for each LED, the total voltage drop of each string differs from the other strings in the array. The forward voltage also depends on the LED diode junction temperature, as previously mentioned. To ensure uniform light output for all LEDs, equal current should be designed to flow through each string of LEDs.

The traditional way to drive a parallel LED array is to connect a current limiting resistor in series with each string and power all the strings using a single voltage source. A substantial voltage drop is required across the resistor to ensure that the current will stay in the desired range, in the presence of temperature and device-to-device voltage variations. This method is inexpensive and suffers from power inefficiency and heat dissipation. It also requires a stable voltage source.

A better way of powering the LED array is to regulate the total current through all the strings and devise a means to divide that total current equally among the LED strings. This is active current control and is the subject of the next section.

3.1.2 Active Current Control

Since a series resistor is not a good current control method, especially when the supply voltage has a wide tolerance, we will now look at active current control. Active current control uses transistors and feedback to regulate the current. Here we will only consider

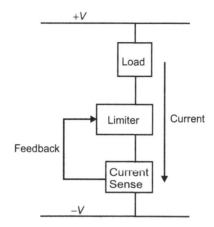

Figure 3.2: Current Limiter Functions.

limiting LED current when the energy is supplied from a voltage source; driving LEDs using energy from current sources will be discussed in Section 3.2.

A current limiter has certain functional elements: a regulating device, such as a MOSFET or bipolar transistor; a current sensor, such as a low value resistor; and some feedback (with or without gain) from the current sensor to the regulating device. Fig. 3.2 shows these functions.

The simplest current limiter is a depletion-mode MOSFET; it has three terminals called gate, drain, and source. Conduction of the drain–source channel is controlled from the gate–source voltage, like any other MOSFET. However, unlike an enhancement MOSFET, a depletion-mode MOSFET is "normally-on" so current flows when the gate–source voltage is zero. As the gate voltage becomes negative with respect to the source, the device turns off, see Fig. 3.3. A typical pinch-off voltage is -2.5 V.

A current limiting circuit with a depletion-mode MOSFET uses a resistor in series with the source to sense the current (Fig. 3.4). The gate is connected to the negative supply (0 V). As current flows through the resistor, the voltage drop across it increases. The voltage at the MOSFET source becomes in higher potential compared to the 0 V rail and the MOSFET

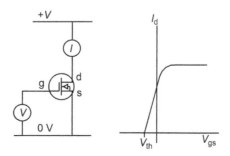

Figure 3.3: Depletion MOSFET Characteristics.

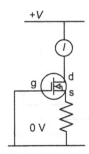

Figure 3.4: Depletion MOSFET Current Limiter.

gate. In other words, compared to the MOSFET source, the gate becomes more negative. At a certain point, when the voltage drop approaches the MOSFET pinch-off voltage, the MOSFET will tend to turn off and thus regulate the current.

The main drawback of using depletion-mode MOSFETs is that the gate threshold voltage (V_{th}) has a wide tolerance. A device with a typical V_{th} of -2.5 V will have threshold range of -1.5 V to -3.5 V. However, the advantage is that high drain–source breakdown voltages are possible and so a limiter designed using a depletion-mode MOSFET can protect against short transients (longer periods of high voltage would tend to overheat the MOSFET).

A simple integrated current limiter is a voltage regulator in the place of the depletion-mode MOSFET, as shown in Fig. 3.5. This uses an internal voltage reference and so tends to be quite accurate. The disadvantage is that there is a minimum dropout voltage, which can be as high as 4 V. This type of current limiter can be used for current sink or current source regulation, depending on whether the load is connected to the positive or negative supply rail.

In Fig. 3.5, the well-known linear voltage regulator LM317 is used as a current limiter. The LM317 has three terminals, labeled "IN," "OUT," and "ADJ." The feedback pin labeled

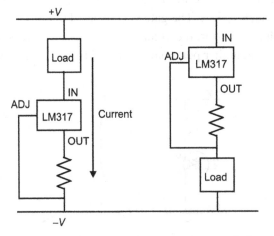

Figure 3.5: Linear Regulator as Current Limiter.

"ADJ" (adjust) controls the regulation of the current. A current sensing resistor is placed between the "OUT" terminal and the load, which is also connected to the "ADJ" terminal. Due to the series connection, the voltage across the current sensing resistor is proportional to the output current and will cause the voltage at the "ADJ" terminal to be lower than the "OUT" terminal. The LM317 feedback limits the current when the "ADJ" terminal voltage reaches 1.25 V below the "OUT" terminal.

If accurate current limiters are used in series with each LED string, parallel strings of LEDs and their limiters can be connected to the same voltage source. In this case, each string will have approximately the same current. With the same current flowing through each LED, the light produced will be almost the same for each LED and thus no "bright spots" will be seen in the LED array.

The current limiters described here are purely to show how LEDs can be driven from a constant voltage supply. Further linear regulators are described in Chapter 4 and switching regulators are described in Chapter 5.

3.1.3 Short Circuit Protection

The current limiting circuits described in the previous section will provide automatic short circuit protection. If the LED goes short circuit, a higher voltage will be placed across the current limiter. Power dissipation is the main issue that needs to be addressed.

If the power dissipation cannot be tolerated when the load goes short circuit, a voltage monitoring circuit will be needed. When a voltage higher than the expected voltage is placed across the current limiter, the current must be reduced to protect the circuit. In the previously described LM317 circuit, the regulator itself has thermal shutdown.

3.1.4 Detecting Failures

If we have a short circuit condition in the LEDs, the voltage across the current limiter will increase. We can use this change to detect a failure. In the circuit shown in Fig. 3.6, a 10-V Zener diode is used in series with the base of an NPN transistor. When the voltage at the "IN" terminal of the LM317 reaches about 11 V, the Zener diode conducts and turns on the transistor. This pulls the "FAILURE" line to 0 V and indicates a short circuit across the LEDs.

3.2 Current Source

Since an LED behaves like a constant voltage load, it can be directly connected to a current source. The voltage across the LED, or string of LEDs, will be determined by the characteristics of the LEDs used. A pure current source will not limit the voltage, so care must be taken to provide some limit; this will be covered in more detail in the next section.

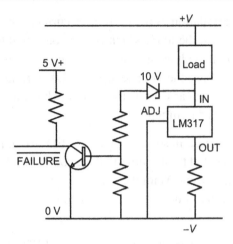

Figure 3.6: Shorted Load Protection.

Parallel strings of LEDs can be driven by current-sharing circuits. The simplest of these is a current mirror, which shares the current equally between strings based on the current flowing through the primary string. Fig. 3.7 shows a simple current mirror. The basic principle relies on the fact that matched transistors will have the same collector current, provided that their base–emitter junctions have the same voltage across them. By connecting all the bases and all the emitters together, every base–emitter junction voltage must be equal and therefore every collector current must be equal.

The primary LED string is the one that controls the current through the other strings. Since the collector and base of transistor Q1 are connected, the transistor will be fully conducting until the collector voltage falls low enough for the base–emitter current to limit. Other transistors (Q2–Qn) have their base connections joined to Q1, and will conduct exactly the

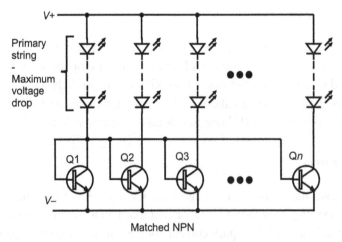

Figure 3.7: Current Mirror.

same collector current as Q1 since the transistors are matched. The total current through Q1 to Q*n* will equal the current source limit.

The voltage drop across the LEDs in the primary string must be higher than any other string in order for the current mirror to work correctly. In the slave strings, some voltage will be dropped across the collector–emitter junction of the transistors Q2–Q*n*. The slave circuits adjust the current by raising or lowering this surplus voltage drop across the transistor.

3.2.1 Self-Adjusting Current Sharing Circuit

As an alternative, the current sharing circuit shown in Fig. 3.8 automatically adjusts for string voltage.

Assuming that the LED array is driven from a current source, there will be equal current division among all connected branches. If any branch is open due to either a failure or no connection by design, the total current will divide evenly among the connected branches. Unlike the simple current mirror, this one automatically adjusts for the maximum expected voltage difference between strings of LEDs, which is a function of the number of LEDs in the string and the type of LED used. The components must be able to dissipate the heat generated by the sum of each string current and the headroom voltage drop across the regulator for that string.

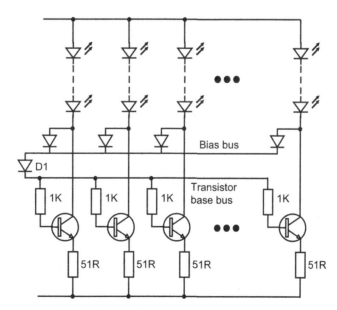

D1 headroom adjust one or more diodes in series

Figure 3.8: Self-Adjusting Current Sharing Circuit.

In high reliability applications, the failure of a single LED should not significantly affect the total light output. The use of the current divider will help the situation. When an LED fails short circuit, the voltage of the string containing the shorted LED will have less voltage. The self-adjusting current sharing circuit will accommodate the change in voltage and still distribute the current equally. When an LED fails open circuit, the current divider will automatically redistribute the total current among the remaining strings, thus maintaining the light output. In this application, an extra diode string can be added for redundancy, so that any single failure will not cause the remaining LEDs to operate in an overcurrent condition.

Equality of current division among the branches is dependent on the close matching of the transistors, which are in close vicinity (ideally a single package with several matched transistors). When any of the transistors saturate due to large variation of the string voltages, equal current division will be lost.

Diodes connected to each collector detect the voltage of each branch. The highest branch voltage (corresponding to the LED string with the lowest forward voltage) is used to bias the transistors in the linear operating region. The cathode of each diode is connected to a common "bias bus."

To accommodate variations in string voltages and keep the current divider transistors from saturation, diodes are connected between the "bias bus" and the "transistor base bus." More than one external diode can be used to accommodate large voltage variations. If the string voltage variation is less than one diode drop, the two buses can be joined.

When a branch is not connected, there will be higher base current flowing in the associated regulating transistor. This could interfere with the current division in the connected branches, so a resistor (about 1 kΩ) is connected from the "transistor base bus" to each transistor base to ensure correct operation of the overall circuit.

3.2.2 Voltage Limiting

In theory, the output voltage of a constant current driver is not limited. The voltage will be the product of the current and load resistance in the case of a linear load. In the case of an LED load, the voltage limit will depend on the number of LEDs in a string. In practice, there will be a maximum output voltage because components in the current source will break down eventually. Limiting the LED string voltage is necessary to prevent circuit damage and the voltage level will depend on the particular circuit. In linear regulators or buck (step-down) switching regulators, the voltage is limited by the source supply voltage. But in boost regulators the output voltage can rise to very high levels and a voltage feedback scheme to provide overvoltage protection should be provided.

Safety regulations will be covered in Chapter 10, but Underwriters Laboratories (UL) Class 2 and safety electrical low voltage requirements limit any potential to 60 V DC, or 42.4 V

AC; so equipment designed to meet these requirements should consider both mains supply isolation (if applicable) and output voltage limiting. The number of LEDs in a string will be restricted in this case, so that the total string voltage remains below 60 V.

3.2.3 Open Circuit Protection

Some constant current drivers, especially switching boost converters, will produce a sufficiently high voltage to destroy the driver circuit. For these types of driver a shutdown mechanism is required. Using a Zener diode to give feedback when the output voltage exceeds a certain limit is one method. Most integrated circuits (ICs) have an internal reference and comparator circuit to provide this function. A potential divider comprising two resistors is usually used to scale down the output voltage to the reference voltage level. Some overvoltage detectors within ICs have a latched output, requiring the power supply to be turned off and then turned on again before LED driver functions are enabled. Other circuits will autorestart when the open circuit condition is removed (i.e., when the LEDs are reconnected).

3.2.4 Detecting LED Failures

In a constant current circuit, a failure of a single LED within a string of series connected LEDs can mean that either a whole string is off (open circuit LED) or a single LED is off (short circuit LED).

In the case of an open circuit LED, the load is removed and so the output voltage from the current source rises. This rise in voltage can be detected and used to signal a failure. In circuits where overvoltage protection is fitted, this can be used to indicate a failure. Similarly, in the case of a short-circuited LED, the output voltage from the current source drops, so the drop in voltage can be used to indicate a failure. In some switching circuits, indirect measurement can be made by detecting a change in the MOSFET gate drive signal (the gate drive pulse width will change as the output voltage changes).

If a current mirror is used to drive an array of LEDs with a number of parallel strings, the result of an open circuit LED will depend on which string the LED is located. In a basic current mirror as shown in Fig. 3.7, a failure in the primary string will cause all the LEDs to have no current flow and not be lit. Detection of the rise in output voltage would be a solution. However, if the failure were in a secondary string, there would be higher current flowing in the other strings and the output voltage would not rise very much (only due to the extra current flowing through the ESR). The voltage at the transistor collector of the broken string would fall to zero since there is no connection to the positive supply and this could be detected.

Another technique, for low current LEDs, is to connect the LED of an optocoupler in series with the LED string. A basic optocoupler has an LED and a phototransistor in the same package. Current through the optocoupler LED causes the phototransistor to conduct. Thus when current is flowing through the LED string and the optocoupler's internal LED, the phototransistor is conducting. If the string goes open circuit, there is no current through the optocoupler's LED and the phototransistor does not conduct.

3.3 Testing LED Drivers

Although testing an LED driver with the actual LED load is necessary, it is wise to use a dummy load first. There are two main reasons for this: (1) cost of an LED, especially for high power devices, can be greater than the driver circuit; and (2) operating high-brightness LEDs for a long time under test conditions can cause eye strain and temporary sight impairment (snow blindness). A further reason is that some dummy loads can be set to limit the current and so enable fault-finding to be made easier.

3.3.1 Zener Diodes as a Dummy Load

Fig. 3.9 shows how Zener diode can be used as a dummy load. This is the simplest and cheapest load. The 1N5334B is a 3.6 V, 5 W Zener diode (3.6 V typical at 350 mA). This is not the perfect dummy load. This reverse voltage is slightly higher than the typical forward voltage of 3.42 V of a Lumileds "Luxeon Star" 1 W LED. The 1N5334B has a dynamic impedance of 2.5 Ω, which is higher than the Luxeon Star's 1-Ω impedance. The impedance will have an effect on some switching LED drivers that have a feedback loop. For simple buck circuits, the impedance only has a small effect.

An active (electronic) load is more precise. A constant voltage load will have (in theory at least) zero impedance, so simply adding a small value series resistor will give the correct impedance. Commercial active loads can be set to have constant current or constant voltage—a constant voltage setting is required to simulate an LED load.

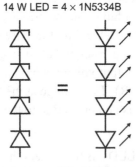

Figure 3.9: Zener Diode Dummy Load.

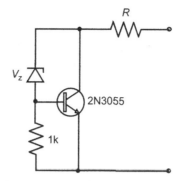

Figure 3.10: Active Dummy Load.

A constant voltage load built using a low-cost discrete solution is shown in Fig. 3.10. This is a self-powered load and so can be isolated from ground. The Zener diode can be selected to give the desired voltage (add 0.7 V for the emitter–base junction of the transistor). The transistor should be a power device, mounted on a heat sink.

The circuit in Fig. 3.10 has low impedance. Although the Zener diode does have a few ohms impedance, the current through it is very small and the effect of the transistor is to reduce the impedance by a factor equal to the gain HFE. Suppose the transistor HFE = 50 at 1 A and the Zener diode impedance $Z_d = 3\ \Omega$. Changing the collector current from 500 mA to 1 A will cause the base current to rise from 10 to 20 mA. A 10 mA change in current through the Zener diode will cause 30 mV voltage rise. This change at the transistor collector is equivalent to an impedance of 30 mV/0.5 A = 0.06 Ω. In other words, the circuit impedance is equal to the Zener diode impedance divided by the transistor gain.

An impedance of 0.06 Ω is unrealistically low, but a power resistor can be added in series to give the desired load impedance. Due to the potentially high load current, both the transistor and series resistor should be rated for high power. The transistor should be mounted on a large heat sink.

3.4 Common Mistakes

The most common mistake is to use expensive high power LEDs when testing a prototype circuit. Instead, a dummy load should be used, using either an electronic load or a number of Zener diodes connected in series. If using Zener diodes, 3.6 V 5 W types should be used in place of each LED. Only once the circuit has been tested under all conditions should LEDs be used.

Another mistake is to connect LEDs in parallel. It can be done if the LEDs are very well matched, both thermally and their forward voltage drops, but this is very difficult to achieve. A current regulator/current mirror in series with each LED string is by far the best option.

3.5 Conclusions

A constant voltage–regulated LED driver is preferred when there are a number of LED modules that can be connected in parallel. Each module will have its own linear or switching current regulator, so the LED current is constant. An example would be channel lighting, as used in shop name boards.

A current-regulated LED driver is preferred when it is desirable to have a number of LEDs connected in series. A series connection ensures that all the LEDs have the same current flowing through them and the light output will be approximately equal. If the load voltage (of the LEDs) is just below the source voltage, a linear current regulator is a good cost-effective solution. A switching current regulator is needed if the LED load voltage is higher than the source voltage, or if the LED load voltage is significantly lower than the source voltage. A switching current regulator is also the favored option while driving high power LEDs, for reasons of efficiency. An efficiency of 90% is normally achievable.

Linear Power Supplies

Chapter Outline

Linear power supplies for driving light-emitting diodes (LEDs) are preferred for a number of reasons. The complete absence of any electromagnetic interference (EMI) radiation is one important technical reason. Low-overall cost is an important commercial reason. However, they also have disadvantages: in some applications they have low efficiency and hence they introduce thermal problems. It may be necessary to add a heat sink, thus increasing cost and size.

Switched linear regulator ICs can be used to good effect when the circuit is powered directly from the AC mains supply. In this case, a number of linear regulators are used to drive a long string of low-current LEDs. The total number of LEDs in series has to be high enough for the forward voltage drop to be close to the peak AC line voltage. In a typical 230-V AC application, about 100 LEDs need to be connected in series. For 120-V AC applications, about 50 LEDs must be connected in series. Switched linear regulators are described in Section 4.3.

4.1 Voltage Regulators

Many voltage regulators are based on the LM317 originally from National Semiconductor, but are now made by a number of manufacturers. Inside the LM317 are: (1) a power switch, which is an NPN transistor; (2) a voltage reference set to produce 1.25 V, and (3) an operational amplifier (op-amp) to control the power switch, as shown in Fig. 4.1. The op-amp

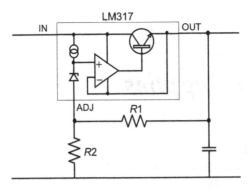

Figure 4.1: LM317 Regulator.
ADJ, Adjust pin; *IN*, input; *OUT*, output; *R*, resistor.

tries to keep the voltage at the output equal to the voltage at the adjust pin (ADJ) minus the reference voltage.

To produce a certain output voltage, a feedback resistor is connected from the output (OUT) to the ADJ pin and a sink resistor is connected from the ADJ pin to ground, thus creating a potential divider. Usually the feedback resistor is set to 240 Ω, to draw a minimum of 5 mA from the regulator and help maintain stability. A capacitor on the output terminal also helps with stability. The output voltage is given by the equation:

$$V_{OUT} = 1.25 \cdot \left(1 + \frac{R2}{R1}\right) + I_{ADJ} \cdot (R2)$$

Note, I_{ADJ} = 100 µA, worst case.

Variations of the LM317 regulator include fixed positive voltage versions (LM78xx) and negative voltage versions (LM79xx), where "xx" indicates the voltage; that is, LM7805 is a +5-V, 1-A regulator.

The LM317 and its variants need a minimum input to output voltage difference to operate correctly. This is typically in the range 1–3 V, depending on the current through the regulator (higher current requires a higher-voltage differential). This input to output voltage difference is equal to the voltage across the internal constant current generator, as the OUT pin is at the same potential as the voltage reference.

Low drop-out voltage regulators sometimes use a PNP transistor as the power switch, with the emitter connected to the input (IN) terminal and the collector connected to the OUT terminal (Fig. 4.2). They also have a ground pin that enables an internal reference voltage to be generated independent of the input to output voltage differential. A drop-out voltage of less than 1 V is possible.

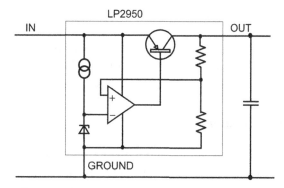

Figure 4.2: Low Drop-Out Voltage Regulator.

4.1.1 Voltage Regulators as Current Source or Sink

In Fig. 4.3 are shown two circuits using a voltage regulator as a current limiter, one is configured as a current source and the other as a current sink.

As previously described, the LM317 regulates when there is +1.25 V between the OUT and ADJ pins. In Fig. 4.3, a current sense resistor ($R1$) is connected between the OUT and ADJ pins. Current flowing through $R1$ will produce a voltage drop, with the OUT pin becoming more positive than the ADJ pin. When the voltage drop across $R1$ reaches 1.25 V, the LM317 will limit the current. Thus the current limit is given by: $I = \dfrac{1.25}{R1}$

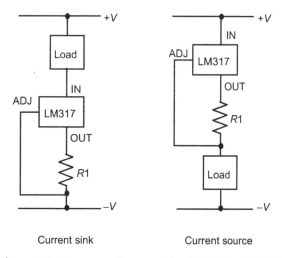

Figure 4.3: Constant Current Circuits Using LM317.

4.2 Constant Current Circuits

There are many constant current circuits; some using integrated circuits, some using discrete components, and others using a combination of both ICs and discrete devices. In this subsection, we will examine some examples of each type.

4.2.1 Discrete Current Regulators

A simple constant current sink uses an op-amp with an input voltage range that extends to the negative rail, as shown in Fig. 4.4. To set the current level, a voltage reference is required. The voltage drop across a current sensing resistor is compared to the reference voltage and the op-amp output voltage rises or falls to control the current. The voltage reference can be a temperature-compensated precision reference, or a Zener diode. A Zener diode generally has a smallest temperature coefficient and lowest dynamic impedance at a breakdown voltage of 6.2 V. In Fig. 4.4, a 3.9-V Zener diode is used because the supply voltage is only 5 V.

4.2.2 Integrated Regulators

There are a number of integrated current regulators suitable for driving LEDs. They can use bipolar or CMOS technology and have a wide range of working voltages. Lower-voltage (typically bipolar) types generally have lower costs, but there are times when high-voltage ratings are needed, particularly for transient suppression. An example where higher voltage is needed is shown in Fig. 4.5.

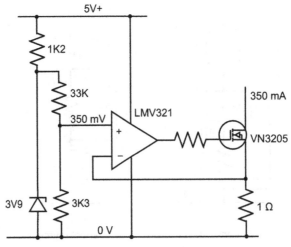

Figure 4.4: Constant Current Sink Using Operational Amplifier (Op-Amp).

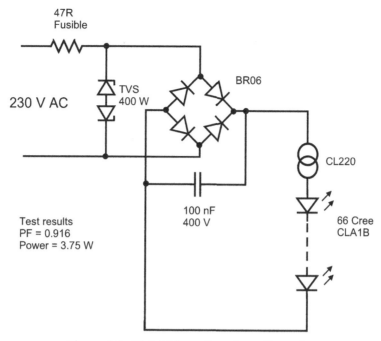

Figure 4.5: High-Voltage Regulator Example.

The Microchip CL220, in the aforementioned example, is a 20-mA current regulator that has a 220-V rating. When used with a high number of LEDs in series, it can be powered from a rectified AC mains supply. The number of LEDs required depends on the AC mains voltage, but for 230-V AC operation about 66 LEDs are needed. The surprising feature is the good power factor (0.916), which is mainly due to the equivalent series resistance of the many low-current LEDs used. The load is thus mainly resistive. Power factor concerns AC mains powered lamps and will be described in Chapter 8, but suffice to say here is that a power factor of 1 is a perfect resistive load and that regulations require a good power factor in most LED lighting applications.

4.3 Switched Linear Current Regulators for AC Mains Operation

Switched linear regulators intended for AC mains powered applications employ multiple current regulators and are for LED lamps where there is a long string of LEDs wired in series. There are tapping points along this LED string for connection to the linear regulators. For efficiency, the total number of LEDs in series has to be high enough, so that the LED string forward voltage drop is just below the peak AC voltage. There are two types of switched linear current regulators. The first type is integrated, like the Microchip CL8800, so that the switching off of each regulator is controlled by subsequent stages. The second type is modular, using a set of cascaded regulators, like the Texas Instruments TPS92410/TPS92411 chipset.

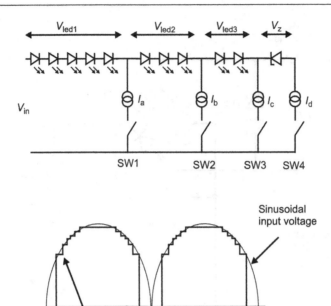

Figure 4.6: Switched Linear Regulator Principle.

4.3.1 Integrated Switched Linear Regulators

The principle of switched linear regulators is illustrated in Fig. 4.6. In this case I am just showing four regulators, for clarity. The current limit of each regulator rises as we go from regulator 1 to regulator 4. Note the Zener diode used instead of LEDs for the last regulator; this provides no light but limits the voltage drop across the current regulator. In practice, the last string of LEDs provides little light, and using a Zener diode prevents low-frequency flicker as the AC voltage peak level varies over time.

A good example of a switched linear regulator IC is the Microchip CL8800, comprising six current regulators (Fig. 4.7). Connected to the positive side of the supply voltage is the LED string, which is tapped at various points along its length. Each tapping point is connected to a CL8800 current regulator input. A current regulator can only conduct when the supply voltage exceeds the forward voltage drop of all the LEDs connected to it. As higher order regulators have more LEDs between them and the supply, the turn-on voltage for the current regulators are progressively increasing as we go from regulator 1 to regulator 6.

The current limits of the six regulators are set individually by external resistors connected, in series, to the 0-V rail. This series connection of the resistors ensures that the current limits

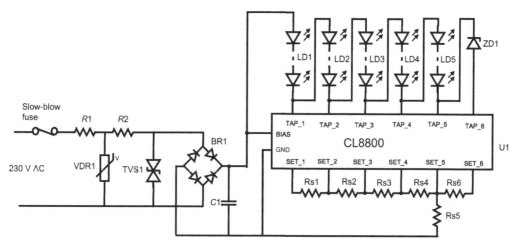

Figure 4.7: Typical CL8800 Application Circuit.

are progressively increased. Regulator 1 has the lowest current limit and regulator 6 has the highest current limit. The regulators are internally coupled, so that when a higher-current regulator starts to conduct, the regulator immediately before it starts to turn off. For example, when current regulator 3 starts to conduct, current regulator 2 starts to turn off. Current regulator 1 is already off and, because of the progressive turn-on voltage, current regulators 4, 5, and 6 are not conducting.

In practice, the number of LEDs in each section of the LED string is progressively reduced. For example, string 1 between the supply voltage and regulator 1 may have 45 LEDs, but string 2 between regulator 1 and regulator 2 may only have 22 LEDs. The last string may have as few as 8 LEDs (or may be replaced by a Zener diode to prevent low-frequency flicker). This arrangement means that the voltage steps, between one regulator turning off and the next taking over, become smaller as the supply voltage increases.

The current steps are not linear either. The current steps are graduated, so they are made large for the first regulator and small for the last regulator. Example current settings are 45 mA for regulator 1, then 60 mA for regulator 2, followed by 80 (reg. 3), 88 (reg. 4), 92 (reg. 5), and finally 95 mA for regulator 6. Note that the schematic shows the current setting resistor connected to ground from terminal set_5, rather than set_6. This is because the reference voltage for the 6th regulator was made higher than the other regulators, for some unknown reason.

The combination of progressively smaller voltage and current steps produced a very high power factor, typically 0.98. The coupling of regulators, so that they are turn on and off gradually, gives a very low-harmonic distortion of 5–15% typically.

4.3.2 *Modular Cascade Linear Regulators*

A good example of a modular cascade linear regulator is the Texas Instruments TPS92410 and TPS92411 chipset (Fig. 4.8). The TPS92410 is a linear controller for regulating the current. The IC uses changes in the rectified AC mains voltage to vary the current limit. The current is thus made proportional to the instantaneous AC mains voltage, so the power factor is high. The TPS92411 is a LED bypass switch, so that some LED strings are shorted out, depending on the AC mains voltage. This allows the efficiency to remain high.

The typical application for 120-V AC operation uses three LED strings in series. The LED strings have to be different lengths, with the first string (80-V forward voltage) being double the length of the second (40-V forward voltage). The second string is double the length of the third string (20-V forward voltage).

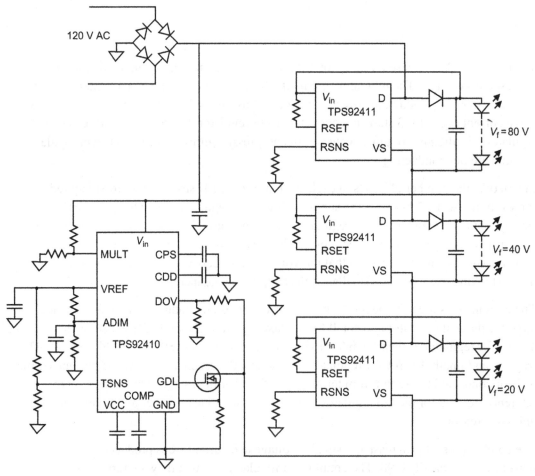

Figure 4.8: Typical TPS92410/TPS92411 Application Circuit.

Switching order: rising supply voltage

Top (80 V)	Middle (40 V)	Bottom (20 V)	V-supply
0	0	0	0 V
0	0	1	20 V+
0	1	0	40 V+
0	1	1	60 V+
1	0	0	80 V+
1	0	1	100 V+
1	1	0	120 V+
1	1	1	140 V+

0 denotes switch ON (LEDs not lit)
1 denotes switch open and LEDs lit

Figure 4.9: Binary Illumination Sequence.

During operation, as the AC mains voltage rises to about 20 V, the first TPS92411 switch opens and the associated LED string is powered. As the voltage rises further, to about 40 V, the second LED is powered and the first TPS92411 switch closes. Now only the second string is lit. As the voltage rises to about 60 V, the first TPS92411 switch opens again and both first and second LED strings are lit. The next step would be for the third LED string to be lit and both the first and second TPS92411 switches closing, thus extinguishing the associated LED strings. The LED strings are lit in a binary sequence, as shown in Fig. 4.9.

4.4 Advantages and Disadvantages

The advantage of linear power supplies is that they produce no EMI radiation. This advantage cannot be overstated, as the cost of materials and development time spent on eliminating EMI can be high.

A switching power supply may appear to have few components, but this does not take into account the EMI filtering and screening. These additional circuits can double the overall cost of the LED driver. If the LEDs are distributed, such as in channel lighting where there is no opportunity to shield any EMI, both common mode and differential filtering are required. And common mode chokes are expensive!

One disadvantage of a linear LED driver can be low efficiency, which is the ratio of the LED voltage to the supply voltage. The efficiency is low only if the supply voltage is somewhat higher than the LED voltage. In these cases, poor inefficiency causes the introduction of thermal problems. A heat sink may be required, which is bulky and moderately expensive. It should be noted that where the supply voltage is only slightly higher than the LED voltage, the efficiency of a circuit using linear regulator could be higher than one using a switching regulator.

Linear mains–powered LED drivers have the disadvantage of large size because a step-down transformer is almost always required (unless the LED string voltage is very near to the peak AC supply voltage). A 50- or 60-Hz mains transformer is bulky and heavy. Smoothing capacitors after the bridge rectifier are also very bulky. The efficiency will vary as the AC supply voltage rises and falls over a long period because the difference between the rectified voltage and the LED string voltage will change.

Switched linear or modular cascaded linear regulators can be used from the AC mains supply. They have the advantage of not needing a transformer, although the disadvantage of not being isolated from dangerously high voltages. They are quite small and surprisingly efficient because most of the voltage drop is across the LED load. However, they do need a considerable number of LEDs in series.

4.5 Limitations

The main limitation of a linear supply is that the LED voltage will always be lower than the supply voltage. Linear voltage and current sources cannot boost the output voltage so that the output is higher than the input. In cases where the output voltage needs to be higher than the input voltage, a switching regulator is necessary. These will be discussed in the next few chapters.

4.6 Common Errors in Designing Linear LED Drivers

The most common error is to ignore the power dissipation. Power dissipation is simply the voltage drop across the regulator multiplied by the current through it. If the voltage drop is high, the current must be limited to stay within the device package power dissipation limits. A surface-mount D-PAK package may be limited to about 1 W, even when there is some copper area soldered to the tab terminal. Heat sinks are now available for surface mount packages, which eases the problem.

Another error is to ignore the start-up conditions. The voltage rating of the regulator must be high enough to allow for the output being connected to 0 V (ground). This is because at start-up, the output capacitor will be uncharged and thus at 0 V. Only after operating for a short period does the output capacitor charge, which reduces the voltage drop across the regulator. The voltage rating of the regulator should always be greater than the maximum input voltage expected.

Buck-Based LED Drivers

Chapter Outline

The first switching LED driver that we will study is the buck converter, which is the simplest of the switching drivers. The buck converter is also known as a step-down converter and is used in applications where the load voltage is less than the supply voltage. In a buck converter circuit, shown in Fig. 5.1, inductor L and an LED are wired in series, with the LED connected to common.

Switch SW1 is used to connect the supply voltage to the inductor, which causes the current to flow through both inductor and LED, so the LED emits light. The inductor limits the rate

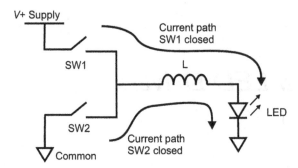

Figure 5.1: Basic Buck LED Driver.

at which the current increases and stores energy during this time. A control circuit will force SW1 to open when the current through the LED reaches a set limit. At this time the inductor has energy stored, which must now be released (an alternative view is that the current in the inductor cannot change instantaneously and so needs to keep flowing until the energy has been spent). A second switch SW2 closes when SW1 opens and connects the inductor to common (also referred to as 0 V or ground), which provides a current path to release the inductor's energy. This secondary current path keeps the LED lit while the switch SW1 is turned off.

As well as storing energy, the inductor L also acts as a low-pass filter to control the rate at which the current rises and falls in the LED. This must be slow enough for the control system to react because the feedback circuit will have a limited response time.

In a practical circuit, switches SW1 and SW2 are semiconductors, either two MOSFETs, or a MOSFET and a diode. Alternate switching of SW1 and SW2 means that the voltage on the switch side of the inductor will be changing from the supply voltage to 0 V (common) and back again, repeated at high frequency. Switching frequencies can typically be in the range 25 kHz–2 MHz. Lower frequencies could be used, but there is a risk of them being in the audible range. Some components have a piezoelectric effect or a magnetorestrictive effect, allowing acoustic energy to be emitted. Switching losses increase with frequency, so high switching frequencies suffer from greater switching losses.

The term "duty cycle" (D) refers to the proportion of time that switch SW1 is on, relative to the whole switching cycle. So if switch SW1 is on for 500 ns and switch SW2 is on for 3500 ns, the switching period is 4000 ns and the duty cycle is 500 ns/4000 ns = 0.125. This can be calculated from the load voltage V_L and supply voltage V_S: $D = V_L/V_S$.

In buck circuits, the maximum duty cycle for reliable operation is usually about 85%. In other words, when switch SW1 is on 85% of the time and off for 15% of the time, this limit means that the output voltage is 85% of the input voltage. So the voltage drop across the inductor, current sense resistor, and the switch cannot be more than 15% of the input voltage. If the input voltage is 5 V, this leaves just $5 \times 0.15 = 750$ mV across the inductor, current sense resistor, and switch.

The 750 mV allowed for voltage drops across other components is quite small. To avoid noisy switching, the voltage drop across the current sense resistor should be greater than 100 mV; 250 mV would be a suitable value. So now the inductor and switch can only have 500 mV dropped across them—an inductor with a low equivalent series resistance (ESR) and a switch with a low on-resistance (R_{ds-on}), are essential.

There are two types of switching: asynchronous and synchronous. In asynchronous switching, switch SW2 is replaced by a diode. This diode switches automatically because when SW1 is turned on, SW2 will be reverse biased (off). When SW1 is turned off, the inductor current wants to keep flowing, to use the inductor's stored energy. The inductor now acts as a voltage source and forward biases the diode at SW2. This diode continues to conduct until the energy stored in the inductor is spent (the current drops to zero), or until switch SW1 turns on again. The diode at SW2 is usually referred to as a "flywheel diode" because it behaves like a flywheel in a motor, circulating the energy.

More details of asynchronous circuits will be given in Sections 5.2–5.4, where control systems are described. A simple schematic of an asynchronous buck circuit is shown in Fig. 5.2.

Here switch SW1 is replaced by MOSFET Q1 and switch SW2 is replaced by diode D1. Note that the positions of the switches are now changed, compared to the diagram in Fig. 5.1, so that Q1 (SW1) is in series with the ground (0 V) rail and flywheel diode D1 (SW2) return current to the positive rail. This transposition is simply to make control simpler, with the feedback and gate drive signals referenced to ground.

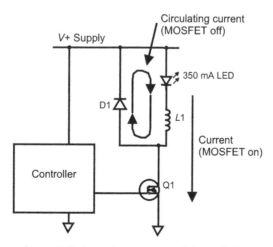

Figure 5.2: Asynchronous Buck LED Driver.

5.1 Synchronous Buck

In low voltage applications, the diode for the return path (at SW2) can be replaced by a second MOSFET. This reduces losses, perhaps allowing 95% efficiency because the voltage drop of a MOSFET during conduction will be much less than a diode. This is known as synchronous switching because the switch timing of the second MOSFET is synchronized with the switch timing of the main MOSFET.

A synchronous buck uses two MOSFET switches at the output, as shown in Fig. 5.3. One switch applies the input voltage to the inductor and the other switch provides the ground connection to the inductor. Only one switch is operated at any time and there is usually a small "dead-time" between switch operations, to give time for the MOSFETs to turn off and thus prevent a momentary short circuit across the supply rails (creating what is known as "shoot-through" current).

During the dead-time, the inherent MOSFET body diode of the ground-connected switch will conduct, to keep the current flowing through the inductor and the LED load. In some applications an ultrafast diode is placed in parallel with the low-side MOSFET because the MOSFET's body diode has a long reverse recovery time. During the reverse recovery time, current can flow in the opposite direction and contribute to switching losses. Bypassing the body diode with a much faster diode lowers the switching losses.

Note that Q1, the high-side switch is shown as an N-channel device. This needs a gate potential several volts higher than the supply voltage to allow the source voltage to reach the maximum level (i.e., the positive supply voltage). A high gate voltage is usually supplied from a bootstrap circuit, like that used in the Linear Technology's LT3743 and LT3592 and ON Semiconductor's NCP5351. The bootstrap circuit is a capacitor charge pump that boosts the supply voltage and stores enough current for the gate drive.

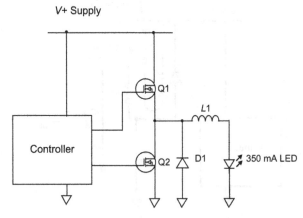

Figure 5.3: Synchronous Buck.

Alternatively, if Q1 were a P-channel MOSFET, this bootstrap voltage would not be needed. The ST device LED2000 is an example of an integrated buck that uses a P-channel MOSFET. Although using an integrated P-channel MOSFET is generally more expensive, with a higher on-resistance/lower current capability, ST's LED2000 is rated at 3 A output and the P-channel switch has just 95 mΩ on-resistance.

A synchronous driver would be essential for very high current loads, such as a laser or car headlights. Laser diodes are also used in video projectors, which have high data rates, and therefore the laser driver needs fast PWM dimming to control the brightness of the rapidly changing image. There are many synchronous driver ICs for these applications. The Texas Instruments (TI) has the TPS92641 with an unusual dimming arrangement—the LED load is shorted by a MOSFET to give 20,000:1 dimming range. Linear Technology has the LT3763 for high current LEDs (up to 20 A), with just 50 mV current sense voltage to maintain efficiency. Fast PWM dimming is achieved in this case by having a MOSFET switch connected in series with the LED load.

For example, suppose that we have a requirement to drive a single white LED at 1 A from a 24 V supply, where the forward voltage drop of the LED is about 3.5 V. Let us first consider using an asynchronous buck circuit, with a Schottky diode as the second switch. The duty cycle is 3.5 V/24 V = 0.146, so the flywheel diode would be conducting for 85.4% of the time, with a forward voltage drop of say 0.5 V, so giving about 12% loss in the diode alone. Now if we use a second MOSFET having a conducting drain–source resistance ($R_{\text{ds-on}}$) of 0.1 Ω, the voltage drop would be 0.1 V, producing about 2.4% loss. Thus we achieve much greater efficiency with the synchronous buck converter.

5.2 Hysteretic Buck

As an alternative to the peak current control buck, hysteretic control can be used in low voltage applications. Hysteretic control uses a fast comparator to drive the MOSFET switch. The input to the comparator is a high-side current sense circuit. In this circuit the voltage drop across a resistor, placed in series with the positive power feed to the LED load, is monitored. This is shown in Fig. 5.4.

During hysteretic control, the MOSFET is turned on when the current level is at or below a minimum reference voltage. The MOSFET is turned off when the current is at or above a maximum reference voltage. This is shown in Fig. 5.5. By this method, the average LED current remains constant, regardless of changes in the supply voltage or LED forward voltage.

The current level is set by a suitable resistor value, given by:

$$R_{\text{sense}} = \frac{1}{2} \cdot \frac{\left(V_{\text{CS(high)}} + V_{\text{CS(low)}}\right)}{I_{\text{LED}}}$$

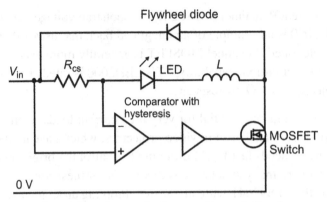

Figure 5.4: Hysteretic Current Control Circuit.

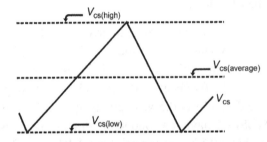

Figure 5.5: Current Sense Voltage (Current in LED Load).

In words, it is the average current sense voltage (midway between the high and low levels) divided by the average LED current required. The upper and lower current sense voltage levels that the comparator uses depend on the particular hysteretic controller being used and are found within the pages of the IC's datasheet.

There are a number of hysteretic buck controllers. The Microchip HV9918 and HV9919/ AT9919 have the same basic functionality, but the HV9918 has a built-in MOSFET, whereas the HV9919/AT9919 drives an external MOSFET and is better for higher current applications. The AT9919 is an automotive version of the HV9919. All three of these ICs have a 200 mV current sense voltage, so a 0.2-Ω resistor is required for sensing 1 A. One advantage of these ICs over those from other suppliers is the analog dimming input, which can be used to input temperature feedback signals from a thermistor that is thermally coupled to the LED load. Having temperature feedback can protect the LEDs, to reduce the power when the LED temperature reaches levels that could reduce the LED lifetime.

The Micrel (now Microchip) MIC3205 is similar to the HV9919, with the same 200 mV current sense voltage. However, it does not have the analog dimming input.

The TI LM3401 uses a different approach; by driving an external P-channel MOSFET it allows a low-side current sense resistor to be connected to ground. Like the other drivers, the current sense voltage is 200 mV. As previously mentioned, using a P-channel MOSFET does have cost and efficiency implications.

Hysteretic control means that the switching frequency is not fixed. The switching frequency can be calculated, but depends on the supply voltage, load voltage, and inductor value being used by the circuit. The frequency is also affected by switching delays (delays in turning the MOSFET on or off, after the upper and lower current limits are reached). Ignoring the delays for the moment, the current rise is given by:

$$E = -L\,\mathrm{d}i/\mathrm{d}t$$

where $E = V_{in} - V_{out}$ (supply voltage − LED voltage drop), L is the inductor value, $\mathrm{d}i$ is the change in current (typically 30% of the average current), and $\mathrm{d}t$ is the rise time. So:

$$T_r = -L\,\mathrm{d}i\,/\,(V_{in} - V_{out})$$

The fall time is:

$$T_f = -L\,\mathrm{d}i\,/\,(V_{out} + V_d)$$

Suppose $L = 22\ \mu\mathrm{H}$, $V_d = 0.6\ \mathrm{V}$, $V_{in} = 12\ \mathrm{V}$, and $V_{out} = 6\ \mathrm{V}$, while $\mathrm{d}i = 0.2\ \mathrm{A}$ (assuming 1 A LED current), then

$$T_r = 22\,\mu\mathrm{H} \times 0.3\,\mathrm{A}\,/\,6\,\mathrm{V} = 1.1\,\mu\mathrm{s}$$

$$T_f = 22\,\mu\mathrm{H} \times 0.3\,\mathrm{A}\,/\,6.6\,\mathrm{V} = 1\,\mu\mathrm{s}$$

The total cycle time is 2.1 μs, so the switching frequency would be 476 kHz.

In reality, the switching delays and the voltage drop across the current sense resistor have a small effect, making the switching frequency lower. For example, in the HV9918 the delay is 70 ns and the voltage drop is 0.2 V. The 0.2 V should be subtracted from V_{in} when calculating the rise time.

$$T_r = 70\,\mathrm{ns} + 22\,\mu\mathrm{H} \times 0.3\,\mathrm{A}\,/\,5.8\,\mathrm{V} = 1.21\,\mu\mathrm{s}$$

$$T_f = 70\,\mathrm{ns} + 22\,\mu\mathrm{H} \times 0.3\,\mathrm{A}\,/\,6.6\,\mathrm{V} = 1.07\,\mu\mathrm{s}$$

Now the total cycle time is 2.28 μs, so the switching frequency becomes 438 kHz. There is some tolerance on this because of tolerances in the value of the inductor, the LED voltage drop (temperature dependent), and the supply voltage.

5.3 Peak Current Control

The Microchip (formerly Supertex) HV9910 integrated circuit was one of the first designed especially for LED driving. A number of semiconductor manufacturers have made copies, or similar designs, based on the HV9910. One such device is the Diodes Inc (formerly Zetex) AL9909 and a similar part is the Clare-Ixys MXHV9910. Peak current control circuits like the HV9910 are suited to high voltage applications because the low-side current sense makes the feedback circuit very simple.

This type of buck driver IC is a good example of a low-cost, low component count solution to implement the continuous mode buck converter (the IC itself needs just three additional components to operate). Linear or PWM dimming can also be easily implemented using the IC. A diagram of the HV9910B is shown in Fig. 5.6.

The HV9910B has two current sense threshold voltages—an internally set 250 mV and an external voltage at the linear dimming (LD) pin. The actual threshold voltage used during switching will be the lower of the two. So if the voltage at the LD pin is 1 V, the internal 250 mV threshold will be used. But if the LD pin voltage is 150 mV, then the threshold voltage is also 150 mV. The low value of sense voltage allows the use of low resistor values for the current sense, which means high efficiency.

So, apart from dimming, there are two main reasons for using the LD pin. In high current applications, a low voltage on the LD pin is used to reduce power dissipation in the current sense resistor. Sometimes the reference voltage can be as low as 100 mV, but with the penalty of accuracy since there can be a 12 mV offset at the current sense comparator input. In

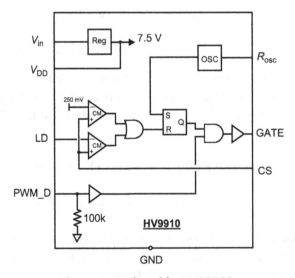

Figure 5.6: Microchip HV9910B.

high accuracy applications, a precision voltage reference can be used to provide a threshold voltage at the LD pin. The precision reference may have 1% tolerance and so may be far superior to the 10% tolerance of the internal reference.

Inside the HV9910, the oscillator timing is controlled by an internal capacitor. This timing capacitor is charged from a current mirror; the other side of the mirror is connected to the R_T pin, which has an external timing resistor connected. When the timing capacitor is charged to a voltage set by an internal reference, a comparator is triggered and this sets a latch to drive the Gate pin high. The latch is reset and the Gate pin pulled low (0 V) when the feedback voltage reaches its threshold level.

If the timing resistor is connected between the R_T pin and ground, a constant current charges the timing capacitor to give a fixed frequency output. Alternatively, if the timing resistor is connected between the R_T pin and the Gate pin, current only flows in the current mirror circuit when the MOSFET is off (Gate pin is 0 V), so now the internal capacitor only charges during the off-time, giving a constant off-time operation. However, in both cases the timing accuracy is not very good; the clock frequency/off-time has 20% tolerance because a silicon oxide layer is used as the dielectric for the internal timing capacitor.

In a simple switching regulator, the oscillator is running continuously. The oscillator triggers a latch, which drives the MOSFET, and the latch is reset by the feedback signal. But there is a deliberate delay in the feedback signal to prevent false triggering, which means that there is a corresponding minimum on-time for the MOSFET. This means that there is always some current in the load. In the case of the HV9910B, the delay and hence minimum on-time is about 280 ns.

Having the MOSFET switched on for this short time causes a proportion of the nominal output current to flow, the proportion is approximately $I_{out} \times 280$ ns/normal on-time. Suppose that the output current of a circuit is normally 350 mA and the on-time is 2800 ns; this is 10 times the minimum on-time. When the LD pin is reduced to 0 V, the minimum load current will be approximately 35 mA, which is 1/10 of the normal current level.

If dimming all the way down to zero is required, the LED string should have a load connected in parallel to draw at least 35 mA. This load could be a resistor, set to draw say 40 mA at the nominal LED voltage, or a constant current diode (such as two CL2 in parallel). In Fig. 5.7, resistor $R3$ is shown as a load to draw the minimum current when the LD pin is at 0 V.

The HV9910B IC operates down to 8 V input, which is required for some applications and can take a maximum of 450 V input, which makes it ideal for AC mains supply applications. The HV9910C requires a higher minimum voltage of 15 V, but this IC has overtemperature protection included.

The HV9910 from Microchip and similar parts from other manufacturers, such as the AL9909 from Diodes Inc, have an internal regulator that supplies 7.5 V to power to the IC's internal circuits from the input voltage, eliminating the need for an external low voltage power supply.

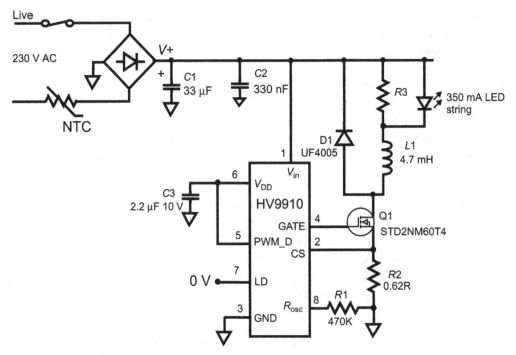

Figure 5.7: Linear Dimming to Zero.

This internal supply should have a good decoupling capacitor, to supply high current pulses when driving the MOSFET gate. In Chapter 11, I discuss multilayer ceramic capacitors and their capacitance variation with both bias voltage and temperature.

The IC is capable of driving the external MOSFET directly, without the need for additional driver circuitry. But care should be taken to consider power dissipation in the linear regulator; when operating from high voltage, I recommend using MOSFETs with low gate charge (Q_g) and a switching frequency of about 50 kHz. The current drawn through the regulator for driving the MOSFET gate will be $I = Q_g \times F_{osc}$. If $Q_g = 30$ nC and $F_{osc} = 50$ kHz, the current will be 1.5 mA, which results in 450 mW dissipation (if the voltage across the regulator is 300 V). Power dissipation can be reduced by adding a high voltage Zener diode in series with the V_{in} pin. This will reduce the voltage across the internal linear regulator and hence reduce the IC's power dissipation.

When operating from low voltages, the limitations on MOSFET gate charge and switching frequency are much reduced. For example, suppose we have a circuit operating from 24 V. We want to know the linear regulator power dissipation due to driving a MOSFET with 50 nC gate charge at 200 kHz. If $Q_g = 50$ nC and $F_{osc} = 200$ kHz, the current due to charging the gate will be 10 mA. Now, the voltage across the regulator is 16.5 V (24 V − 7.5 V V_{DD}), which results in 165 mW dissipation in the regulator circuit.

Note that according to the HV9910 datasheet (Microchip), its linear regulator is only guaranteed to supply 5 mA current if operated from an 8 V supply. This is due to the combination of internal voltage drops and undervoltage protection on V_{DD}. In my experience, more current can be provided with higher voltage supplies, up to about 15 mA. But this is not specified in the datasheet and so the designer is advised to check the limits in his own laboratory!

A circuit operating at fixed frequency, with current mode control, will be unstable if the duty cycle exceeds 50%. In the case of a buck circuit, this means instability if the load voltage exceeds 50% of the supply voltage. If the controller IC can only operate in fixed frequency mode, or fixed frequency is required for another reason, slope compensation is required. Slope compensation is described in Chapter 10. A solution to the stability problem with HV9910 type circuits is to operate in constant off-time mode. This is achieved by connecting the timing resistor between the R_T and Gate pins, as described earlier.

5.4 Average Current Control

Hysteretic current control is an average current control scheme, as described in Section 5.2. In this section, additional average current control schemes are described. These typically use peak current sensing initially, with a "correction" to the switch on-time being applied on subsequent switching cycles. These type of control schemes allow the use of N-channel MOSFET switches and low-side current sensing, which is simple and low cost.

The ON Semiconductor (ON-Semi) NCL30160 operates with a 6.3–40 V input voltage range, using an internal N-channel MOSFET. Initial operation uses a peak current sensing buck operation, where the current through the MOSFET is monitored and the MOSFET is turned off when the voltage across the current sense resistor reaches 220 mV. The MOSFET is turned on again after a time interval set by an internal timer. However, unlike peak current control, in this IC the current is also measured at the moment of turn-on. The timer is then adjusted until subsequent cycles result in 180 mV being detected at turn-on across the current sense resistor. Thus the average current sense voltage is 200 mV (like that of many hysteretic controllers).

The Microchip HV9961 and HV9861A are high voltage LED driver ICs with an input voltage up to 450 V, thus allowing operation from rectified AC mains supplies or the high voltage DC output from a power factor correction circuit (see Chapter 8). These ICs have an output for driving an external MOSFET switch and a comparator input from a low-side current sense resistor. Constant off-time switching is used, so that the circuit is stable for all loads, even those where the load voltage is more than half the supply voltage (duty cycles greater than 50%).

The first switching cycle of the HV9961 type devices uses peak current control, with the peak amplitude set by an internal reference voltage (270 mV). After a fixed off-time, set by an external timing resistor, the MOSFET is turned on and the initial current level is measured. An average current control circuit then adjusts the threshold for the peak current level, so that the second peak is higher and the average current is then at the desired level (the average output current results in an average of 270 mV across the sense resistor).

Subsequent switching cycles continue to adjust the threshold, if required by changes in the operating conditions, to maintain the correct average current. A change in operating conditions could be due to (among other things): the change in input voltage; or the drop in load voltage as the LEDs get warmer; or the change in inductance value with temperature.

Apart from the advantage of accurate current control, the HV9961 and HV9861A also have short circuit protection. This avoids the problem seen with the HV9910 type circuit, where an output short circuit leads to current escalation in the inductor until it saturates, causing the external MOSFET and the control IC to be destroyed.

5.5 Microcontroller-Based Systems

The latest microcontrollers are far more than simple programmable logic controllers. Many have a number of analog functions (peripherals) included and, still better, these functions can be operated independent of the microcontroller core. This means that functions like PWM controllers, analog-to-digital converters, comparators, etc. can continue working regardless of the microcontrollers programming. The detailed design of power supplies for LED driving, using a microcontroller, is beyond the scope of this book; this topic warrants a book on its own. In this section I will give examples of microcontrollers and describe how they are suited for this application.

One concern with microcontrollers is the speed of the peripherals. For example, the comparator built into the PIC16F18313 has maximum response times of 600 ns rising edge and 500 ns falling edge. This speed does allow its use as a current sense comparator, but not for high frequency switching or where the duty cycle is very small.

A second concern is the ability to drive the gate of an external MOSFET. In LED driver ICs, we can find many parts that have gate drive currents of 0.25–2 A. However, if a microcontroller is used, a separate gate drive IC would probably be required. In the case of the PIC16F18313, the absolute maximum I/O current is 50 mA. In addition, the output voltage will be at V_{CC} of the microcontroller (typically in the range 2.5–5 V), which may be insufficient.

The Texas Instruments MSP430G2231 microcontroller has been used in LED driver circuits. The application note SLAA604 shows that the V_{CC} is 6 V in this case and an external MOSFET gate drive circuit is required. The design given in the application note uses a switching speed of 200 kHz, however this is for a circuit with a duty cycle of 0.55, so the on and off periods are almost equal. If the duty cycle was much less, the switching frequency would have to be reduced to allow for the feedback response time.

5.6 Buck Circuits for Low–Medium Voltage Applications

In very low voltage applications, synchronous buck regulators (see Section 5.1) and controller circuits using hysteretic control (see Section 5.2) can be used. For applications where the supply voltage is still low, but exceeds the voltage rating of these controller ICs, we may need to use peak current control ICs. For these applications the circuit shown in Fig. 5.8 can be used.

This circuit shows the Microchip HV9910B being used, but any similar controller IC can be used. Note that the HV9910B will work from an 8 V supply voltage, but the HV9910C (and some others) have a higher minimum supply voltage requirement.

5.6.1 Target Specification

$$\text{Input voltage} = 10\,\text{V} - 30\,\text{V}$$

$$\text{LED string voltage} = 4 - 8\,\text{V}$$

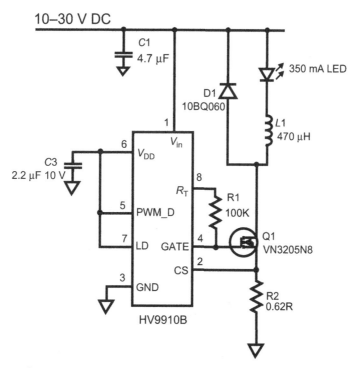

Figure 5.8: Buck Converter for Low Voltage Applications.

$$LED\ current = 350\ mA$$

$$Expected\ efficiency = 90\%$$

5.6.2 Choosing the Switching Frequency and Resistor (R1)

The switching frequency determines the size of the inductor $L1$. A larger switching frequency will result in a smaller inductor, but will increase the switching losses in the circuit. A typical switching frequency for low input voltage applications is: $f_s = 150$ kHz, which is a good compromise. From the HV9910B datasheet, the timing resistor between the R_T pin and ground that is needed to achieve this frequency is 150 kΩ.

However, in this case the maximum output voltage is only 80% of the minimum input voltage. In a buck converter, the duty cycle of the MOSFET switch (proportion of the time that the switch is turned on) will also be 80%. However, in a continuous current regulator, instability will result when the duty cycle goes over 50%. To prevent instability, it is necessary to operate in constant off-time mode. This is achieved with the HV9910 circuit by connecting the timing resistor between the R_T pin and the Gate pin, as previously described.

If we choose a timing resistor that gives a constant off-time of say 5 μs, with an 80% duty cycle the on-time will be 20 μs. The switching frequency will be 40 kHz. At the other extreme, with a 30 V supply and a 4 V load, the duty cycle will be just 13.33%, so the on-time will be 767 ns. Now the switching frequency is 173.4 kHz. The average switching frequency will be about 100 kHz, so we the selection of other components can be based on this. The timing resistor to give 5 μs off-time will be 100 kΩ.

5.6.3 Choosing the Input Capacitor (C1)

An electrolytic capacitor is good to hold the voltage, but the large ESR of these capacitors makes it unsuitable to absorb the high frequency ripple current generated by the buck converter. Thus, metalized polypropylene capacitors or ceramic capacitors in parallel are needed to absorb the high frequency ripple current. The required high frequency capacitance can be computed as

$$C1 = \frac{I_o \times T_{off}}{(0.05 \times V_{min})}$$

In this design example, the high frequency capacitance required is about 4.7 μF, 50 V. This capacitor should be located close to the inductor $L1$ and MOSFET switch Q1, to keep the high frequency loop current within a small area on the PCB. In practice, two such capacitors with a small inductor between them (to make a PI filter) are needed to limit EMI emissions.

5.6.4 Choosing the Inductor (L1)

The inductor value we use depends on the allowed level of ripple current in the LEDs. Assume that ±15% ripple (a total of 30% peak to peak) is acceptable in the LED current.

The familiar equation for an inductor is $E = L \times \dfrac{di}{dt}$. Considering the time when the MOSFET switch is off, so that the inductor is supplying energy to the LEDs, $E = V_{LED} = V_{o,max} = L \times \dfrac{di}{dt}$. Another way of writing this is $L = V_{o,max} \times \dfrac{dt}{di}$. Here, di is the ripple current $= 0.3 \times I_{o,max}$ and dt is the off-time.

Then, the inductor $L1$ can be computed at the rectified value of the nominal input voltage as

$$L1 = \frac{V_{o,max} \times T_{off}}{0.3 \times I_{o,max}}$$

In this example, $L1 = 380\ \mu H$ and the nearest standard value is $470\ \mu H$. Since this value is a little higher than the calculated value, the ripple current will be less than 30%.

The peak current rating of the inductor will be 350 mA plus 15% ripple:

$$i_p = 0.35 \times 1.15 = 0.4\ A$$

The RMS current through the inductor will be the same as the average current (i.e., 350 mA).

5.6.5 Choosing the MOSFET (Q1) and Diode (D2)

The peak voltage seen by the MOSFET is equal to the maximum input voltage. Using a 50% safety rating,

$$V_{FET} = 1.5 \times 30\,V = 45\,V$$

The maximum RMS current through the MOSFET depends on the maximum duty cycle, which is 80% in our example. Hence, the current rating of the MOSFET is

$$I_{FET} \approx I_{o,max} \times 0.8 = 0.28\ A$$

Typically a MOSFET with about 3 times the current is chosen to minimize the resistive losses in the switch. For this application, choose a 50 V, >1 A MOSFET; a suitable device is a VN3205N8, rated at 50 V, 1.5 A.

The peak voltage rating of the diode is the same as the MOSFET. Hence,

$$V_{\text{diode}} = V_{\text{FET}} = 45\,\text{V}$$

The average current through the diode under worst case conditions (minimum duty cycle) is

$$I_{\text{diode}} = 0.87 \times I_{\text{o,max}} = 0.305\,\text{A}$$

Choose a 60 V, 1 A Schottky diode. The International Rectifier 10BQ060 is a suitable type.

5.6.6 Choosing the Sense Resistor (R2)

The sense resistor value is given by

$$R2 = \frac{0.25}{1.15 \times I_{\text{o,max}}}$$

This is true if the internal voltage threshold of 0.25 V is being used. Otherwise, substitute the voltage at the LD pin instead of the 0.25 V into the equation.

For this design, $R2 = 0.625\,\Omega$. The nearest standard value is $R2 = 0.62\,\Omega$.

Note that capacitor $C3$ is a bypass capacitor for holding up the HV9910 internal supply V_{DD} during MOSFET switching, when high frequency current pulses are required for charging the gate. A typical value for $C3$ of 2.2 µF, 16 V is recommended, although in this design the MOSFET gate charge is very low, so a 1 µF, 16 V can be used instead.

5.7 Buck Circuits for High Voltage Input

I will now discuss the design of a high voltage input buck-based LED driver using the HV9910C, or similar type from other manufacturers, with the help of an AC mains input application example. The same procedure can be used to design LED drivers with other input voltage ranges. The schematic is shown in Fig. 5.9.

Designs for an AC input have two problem areas to address. The first is to consider power factor. Power factor correction is described in Chapter 8. Second, in addition to considering the LED driving aspects, we must also consider the low frequency and high voltage issues. As we are applying a low frequency sinusoidal high voltage supply, high value input capacitors are needed to hold up the supply voltage during the cusps between each half-cycle of the input. Applying high voltage across high value capacitors creates a large inrush current that can cause damage, so an inrush limiter (negative temperature coefficient thermistor) is required.

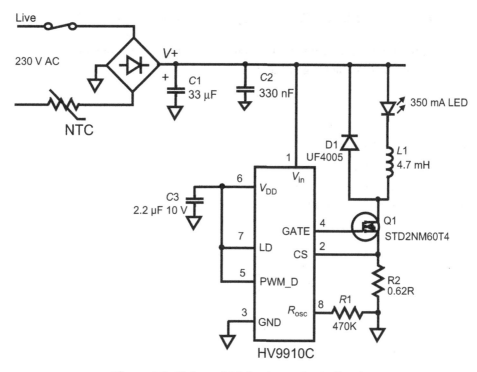

Figure 5.9: Universal Mains Input Buck Circuit.

5.7.1 Target Specification

Ignoring power factor for now, as an exercise we will create a design based on a universal AC input.

$$\text{Input voltage} = 90\,\text{V to } 265\,\text{V AC (nominal } 230\,\text{V AC)}$$

$$\text{LED string voltage} = 20 - 40\,\text{V}$$

$$\text{LED current} = 350\,\text{mA}$$

$$\text{Expected efficiency} = 90\%$$

5.7.2 Choosing the Switching Frequency and Resistor (R1)

The switching frequency determines the size of the inductor $L1$. A larger switching frequency will result in a smaller inductor, but will increase the switching losses in the circuit. A typical switching frequency for high input voltage applications is $f_s = 80\,\text{kHz}$, which is a good compromise. From the HV9910 datasheet, the timing resistor needed to achieve this is $470\,\text{k}\Omega$.

5.7.3 Choosing the Input Diode Bridge (D1) and the Thermistor (NTC)

The voltage rating of the diode bridge will depend on the maximum value of the input voltage. A 1.5 multiplication factor gives a 50% safety margin.

$$V_{bridge} = 1.5 \times \left(\sqrt{2} \times V_{max,ac} \right) = 562 \, V$$

The current rating will depend on the highest average current drawn by the converter, which is at minimum input voltage (DC level allowing for a "droop" across the input capacitor) and at maximum output power. The minimum input voltage must be less than half maximum LED string voltage. For this example, the minimum rectified voltage should be

$$V_{min,dc} = 2 \times V_{o,max} = 80 \, V$$

$$I_{bridge} = \frac{V_{o,max} \times I_{o,max}}{V_{min,dc} \times \eta} = \frac{14}{72} = 0.194 \, A$$

For this design, using a 230 V AC supply, choose a 600 V, 1 A diode bridge.

The thermistor should limit the inrush current to not more than 5 times the steady state current, assuming maximum voltage is applied. The required cold resistance is:

$$R_{cold} = \frac{\sqrt{2} \times V_{max,ac}}{5 \times I_{bridge}}$$

This gives us a 380 Ω resistance at 25°C. Choose a thermistor whose resistance is around 380 Ω and RMS current greater than 0.2 A.

5.7.4 Choosing the Input Capacitors (C1 and C2)

The first design criterion to meet is that the maximum LED string voltage must be less than half the minimum input voltage. As we have already seen, the minimum rectified voltage should be

$$V_{min,dc} = 2 \times V_{o,max} = 80 \, V$$

The hold-up capacitor required at the output of the diode bridge will have to be calculated at the minimum AC input voltage. The capacitor can be calculated as

$$C1 \geq \frac{V_{o,max} \times I_{o,max}}{\left(2 \times V_{min,ac}^2 - V_{min,dc}^2 \right) \times \eta \times freq}$$

In this example,

$$C1 \geq 26.45\,\mu F$$

The voltage rating of the capacitor should be more than the peak input voltage.

$$V_{max,cap} \geq \sqrt{2} \times V_{max,ac}$$

$$\Rightarrow V_{max,cap} \geq 375\,V$$

Choose a 400 V, 33 µF electrolytic capacitor.

The electrolytic capacitor is good to hold the voltage, but the large ESR of these capacitors makes it unsuitable to absorb the high frequency ripple current generated by the buck converter. Thus, a metalized polypropylene capacitor is needed in parallel with the electrolytic capacitor to absorb the high frequency ripple current. The required high frequency capacitance can be computed as

$$C2 = \frac{I_{o,max} \times 0.25}{f_s \times \left(0.05 \times V_{min,dc}\right)}$$

In this design example, the high frequency capacitance required is about 0.33 µF, 400 V. This capacitor should be located close to the inductor $L1$ and MOSFET switch Q1, to keep the high frequency loop current within a small area on the PCB.

5.7.5 Choosing the Inductor (L1)

The inductor value we use depends on the allowed level of ripple current in the LEDs. Assume that ±15% ripple (a total of 30%) is acceptable in the LED current.

The familiar equation for an inductor is $E = L \times \dfrac{di}{dt}$. Considering the time when the MOSFET switch is off, so that the inductor is supplying energy to the LEDs, $E = V_{LED} = V_{o,max} = L \times \dfrac{di}{dt}$. Another way of writing this is $L = V_{o,max} \times \dfrac{dt}{di}$. Here, di is the ripple current $= 0.3 \times I_{o,max}$ and dt is the off-time $dt = \dfrac{\left(1 - \dfrac{V_{o,max}}{\sqrt{2} \times V_{ac,nom}}\right)}{f_s}$. Note, a buck circuit duty cycle is given by $D = \dfrac{V_{out}}{V_{in}}$, so the off-time is $dt = \dfrac{(1 - D)}{f_s}$.

Then, the inductor $L1$ can be computed at the rectified value of the nominal input voltage as

$$L1 = \frac{V_{o,max} \times \left(1 - \dfrac{V_{o,max}}{\sqrt{2} \times V_{ac,nom}}\right)}{0.3 \times I_{o,max} \times f_s}$$

In this example, $L1 = 4.2$ mH. The nearest standard value is 4.7 mH. Since this value is a little higher than the calculated value, the ripple current will be less than 30%.

The peak current rating of the inductor will be 350 mA plus 15% ripple:

$$i_p = 0.35 \times 1.15 = 0.4 \text{ A}$$

The RMS current through the inductor will be the same as the average current (i.e., 350 mA).

5.7.6 Choosing the MOSFET (Q1) and Diode (D2)

The peak voltage seen by the MOSFET is equal to the maximum input voltage. Using a 50% safety rating,

$$V_{FET} = 1.5 \times \left(\sqrt{2} \times 265\right) = 562 \text{ V}$$

The maximum RMS current through the MOSFET depends on the maximum duty cycle, which is 50% by design. Hence, the current rating of the MOSFET is

$$I_{FET} \approx I_{o,max} \times \sqrt{0.5} = 0.247 \text{ A}$$

Typically a MOSFET with about 3 times the current is chosen to minimize the resistive losses in the switch. For this application, choose a 600 V, >1 A MOSFET; a suitable device is an ST part, STD2NM60, rated at 600 V, 2 A.

The peak voltage rating of the diode is the same as the MOSFET. Hence,

$$V_{diode} = V_{FET} = 562 \text{ V}$$

The average current through the diode is

$$I_{diode} = 0.5 \times I_{o,max} = 0.175 \text{ A}$$

The reverse recovery time (T_{rr}) of the flywheel diode is critical. Choose a 600 V, 1 A ultrafast diode. The UF4005 is a low-cost ultrafast type, but the reverse recovery time is 75 ns. For greater efficiency, a faster diode like STTH1R06 ($T_{rr} = 25$ ns) should be used.

Note that in high current, high voltage, buck circuits (typically where the LED current is above 700 mA) it may be necessary to use a high voltage Schottky flywheel diode. This is normally referred to as a silicon carbide (SiC) diode. It is more expensive than an ultrafast silicon junction diode, but it is effective at reducing switching losses; efficiency is increased and heat loss reduced. I have seen a streetlight application where the driver circuit was overheating, but ran cool after a SiC diode was used in place of a silicon junction diode.

5.7.7 Choosing the Sense Resistor (R2)

The sense resistor value is given by

$$R2 = \frac{0.25}{1.15 \times I_{o,max}}$$

This is true if the internal voltage threshold of 0.25 V is being used. Otherwise, substitute the voltage at the LD pin instead of the 0.25 V into the equation. The value of 1.15 in the denominator of the equation allows for the peak current to be 15% above the average current (30% ripple = ±15%). In practice, delays in the HV9910C current sense comparator will cause the peak current to be higher than this.

For this design, $R2 = 0.625\ \Omega$. The nearest standard value is $R2 = 0.62\ \Omega$.

Note that capacitor C3 is a bypass capacitor for holding up the HV9910C internal supply V_{DD} during MOSFET switching, when high frequency current pulses are required for charging the gate. A typical value for C3 of 2.2 μF, 16 V is recommended.

5.7.8 Current Sense Delay

Sometimes it is necessary to add a short delay to the current sense signal. Fig. 5.10 shows a 2K2 Ω resistor and a 100 pF capacitor between the current sense resistor and the CS input at the HV9910C. This delay circuit is often required if the supply voltage is high, to prevent false triggering of the current sense comparator due to the surge current that occurs every time the MOSFET switches on. A high surge current is most likely if the inductor has high parasitic capacitance, typical of a high value power inductor, or if the reverse recovery time of the flywheel diode is slow enough to allow a high reverse recovery current surge.

5.8 AC Circuits With Triac Dimmers

An LED driver powered by an AC triac dimmer needs special additional circuits. These additional circuits are required because of the triac dimmer circuit. Triac dimmers usually use a triac activated by a passive phase shift circuit. Due to switching transients, which would otherwise cause serious EMI problems, the triac is bypassed by a capacitor (typically 10 nF) and has an inductor in series with its output. The triac dimmer circuit is shown in Fig. 5.11.

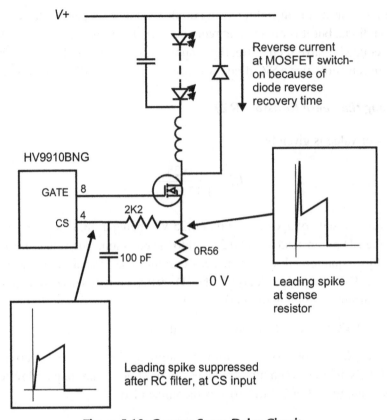

Figure 5.10: Current Sense Delay Circuit.

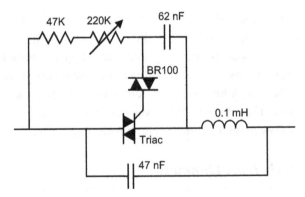

Figure 5.11: Triac Dimmer Circuit.

The input of an inactive LED driver is high impedance, with a large capacitor on the DC side of the bridge rectifier. The capacitor across the triac allows a small current to flow through the bridge rectifier and the smoothing capacitor starts to charge. When the voltage builds up, the LED driver will try to operate. The result is an occasional flicker of the LED.

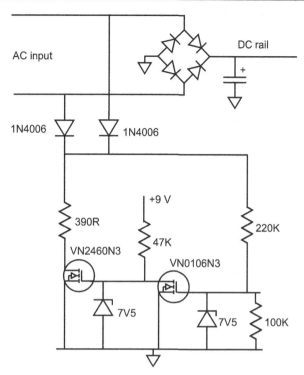

Figure 5.12: Smoothing Capacitor Discharge Circuit.

What is required is a discharge circuit, to keep the smoothing capacitor voltage below that required to start the LED driver. A 390-Ω resistor was found to keep the smoothing capacitor voltage below 5 V. To prevent high power loss when the circuit is active, a simple voltage detector can be used to disconnect the 390-Ω resistor when a voltage above about 8 V is detected. This circuit is shown in Fig. 5.12.

The triac needs to see a load. Once a triac is triggered, it is the load current that keeps it switched on; the triac is a self-sustaining switch. However, an LED driver provides no load until the input voltage has risen above the LED voltage, and it takes a little time for this current to be stable at sufficiently high level to keep the triac turned on. For this reason, an additional load must be switched across the LED driver input at low voltages.

Tests have shown that a 2K2-Ω resistor works as a triac load and that it should remain in circuit until the supply voltage has risen to about 100 V, but should then be switched off until the rising edge of the next half-wave. A latching circuit to provide this function is shown in Fig. 5.13.

These circuits can be combined. The voltage detector for the smoothing capacitor discharge circuit can also be used to provide an enable signal for the LED driver (PWM

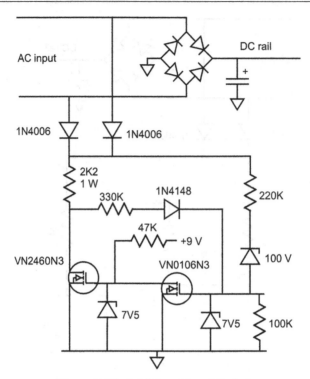

Figure 5.13: Additional Load Switch.

input). Thus when the triac is off, the LED driver is also off. The combined circuit is shown in Fig. 5.14.

5.9 Double Buck

The double buck is an unusual design, as shown in Fig. 5.15. It uses one MOSFET switch, but two inductors (*L2* and *L3*) in series. Diodes steer the current in *L2*, which must operate in discontinuous conduction mode (DCM) for correct operation.

The double buck is used when the output voltage is very low and the input voltage is high. An example is driving a single power LED from an AC supply line. A single buck stage cannot work easily because the on-time of the buck converter is too small, unless a very low switching frequency is used. One disadvantage of the double-buck circuit is that neither side of the LED is connected to a power rail. This means that during switching there is a high common-mode voltage on the LED (the voltage on both sides of the LED rises and falls at high frequency, relative to the power rails). This common-mode signal can generate high levels of EMI radiation.

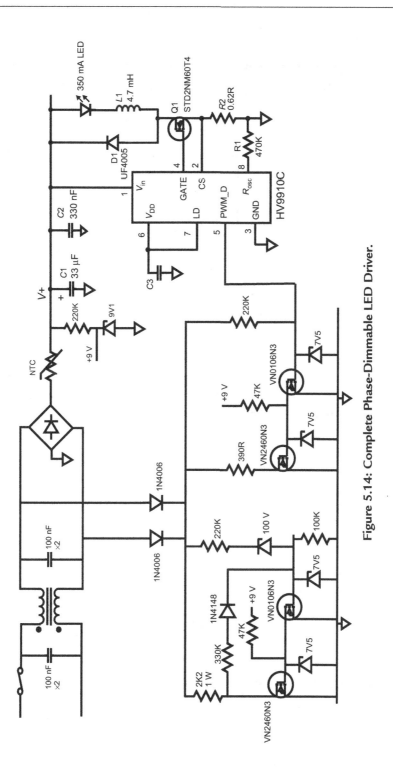

Figure 5.14: Complete Phase-Dimmable LED Driver.

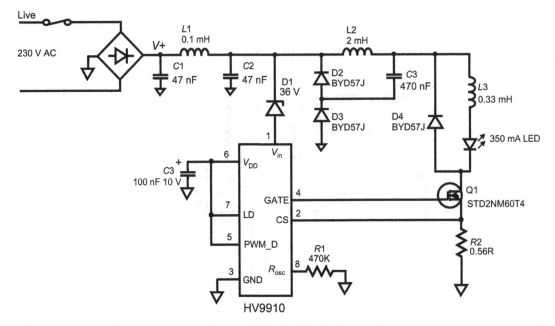

Figure 5.15: Double Buck.

Assume the maximum duty cycle, D_{max}, is less than 0.5; also assume that the first stage ($L2$) is in boundary conduction mode (BCM) at D_{max}. BCM means that the current through the inductor only just falls to zero and the next switching cycle begins.

$$V_{in\,min} = \frac{V_o}{D_{max}^2}$$

or transposed, this becomes:

$$D_{max} = \sqrt{\frac{V_o}{V_{in\,min}}}$$

This assumes that $L2$ is in BCM and $L3$ is in continuous conduction mode (CCM); at the minimum operating input voltage ($V_{in\,min}$).

The storage capacitor voltage at $V_{in\,min}$ and D_{max} is given by the equation:

$$V_{c\,min} = V_{in\,min} * D_{max}$$

The peak current through the input stage inductor, at $V_{in\,min}$ equals:

$$I_{L2\_pk} = 2 * I_{L2\_avg}$$
$$= 2 * \frac{V_o * I_o}{V_{c\,min}}$$

Thus the primary stage inductor $L2$ has a value given by:

$$L2 = \frac{\left(V_{in\,min} - V_{c\,min}\right) * D_{max} * T_s}{I_{L2\_pk}}$$

The transfer ratio for a DCM buck converter (where R is load resistor seen by the converter), is given by:

$$\frac{V_c}{V_{in}} = \frac{2}{1 + \sqrt{1 + \dfrac{8 \times L2}{R \times T_s \times D^2}}}$$

The resistor R seen by the first stage (and assuming second stage is in CCM) is given by:

$$R = \frac{V_c^2}{P_o}$$

$$\Rightarrow R \times D^2 = \frac{\left(V_c \times D\right)^2}{P_o} = \frac{V_o^2}{P_o}$$

Combining the previous two equations (which turn out to be a constant):

$$\frac{V_c}{V_{in}} = K = \frac{2}{1 + \sqrt{1 + \dfrac{8 \times L2 \times P_o}{T_s \times V_o^2}}}$$

we find that D is inversely proportional to V_{in}:

$$D = \frac{V_o}{V_c} = \frac{V_o}{K \times V_{in}}$$

And we can now show that the peak inductor current through $L2$ is a constant over the operating input voltage:

$$\text{Setting } D = K'/V_{in}, K' = V_o / K$$

K' is a constant, since V_o is constant.

$$i_{L2,pk} = \frac{\left(V_{in} - V_c\right) \times D \times T_s}{L2}$$

$$= \frac{V_{in} \times \left(1 - K\right) \times \dfrac{K'}{V_{in}} \times T_s}{L1}$$

$$= \frac{\left(1 - K\right) \times K' \times T_s}{L2}$$

We can now define the average input voltage as the maximum input voltage $\left(\sqrt{2}V_{ac\,max}\right)$ and the minimum operating input voltage:

$$V_{in\,avg} = \frac{(V_{in\,max} + V_{in\,min})}{2}$$

The storage capacitor value is computed based on 10% voltage ripple on the capacitor at $V_{in\,min}$ and D_{max}:

$$C = \frac{0.5 * I_{L2\_pk} * (1 - D_{max}) * T_s}{0.1 * V_{c\,min}}$$

The voltage across the storage capacitor, with average voltage input, is given by:

$$V_{c\,avg} = K * V_{in\,avg}$$

We can now compute the average duty cycle (at average input voltage):

$$D_{avg} = \frac{V_o}{V_{c\,avg}}$$

Computing the value of L3:

$$L3 = \frac{(V_{c\,avg} - V_o) * D_{avg} * T_s}{\Delta I_{L3}}$$

5.10 Buck Design Mistakes

5.10.1 Common Errors in Low Voltage Buck Design

1. Using an inductor that has too high inductance.
    Although increasing the inductor value may seem to be the answer to reduce current ripple, it actually causes problems because the current does not fall enough between switching cycles for proper control by the controller IC. The voltage seen across the current sense resistor at switch-on will be almost at the current sense comparator reference voltage. At switch-on there will be a current surge, caused by the flywheel diode reverse current and the current through the inductor's parasitic capacitance. The smallest current surge will create a voltage spike across the current sense resistor and hence the current sense comparator will trip. This means that the MOSFET will switch off almost immediately after switch-on.
    A typical switching pattern is one proper switching cycle, where energy is stored in the inductor, followed by one short switching pulse. This switching pulse provides very little

energy to the inductor, but generates high switching losses. The result is a less efficient circuit that could suffer from overheating and EMI problems

2. Using the wrong type of flywheel diode.

   A Schottky diode has a low forward voltage drop, which will give low power dissipation. However, in low duty cycle applications the LED current is flowing in the flywheel diode most of the time. A forward voltage of say 0.45 V at 350 mA results in 157.5 mW conduction losses, so an SMA size package works well, but for higher current applications a large SMB or SMC package should be considered. Note that the forward voltage drop of Schottky diodes increases with their current rating, so a 30 V Schottky has much lower V_f than a 100 V Schottky.

5.10.2 Common Errors in AC Input Buck Circuits

1. A common error is trying to drive a single LED from the AC mains supply.

   The duty cycle is V_{out}/V_{in}, so for universal AC input 90–265 V AC, the rectified voltage is about 100–375 V. The worst case is the higher voltage; consider driving a white LED with 3.5 V forward voltage. The duty cycle will be 3.5/375 = 0.9333% duty cycle. If the switching frequency is 50 kHz, with 0.02 ms period, the MOSFET on-time will be just 186 ns. This time is too short for the current sense circuit to react; it needs to be at least 300 ns. Operating at 20 kHz will give an on-time of 466 ns, which is near to the limit for accurate control. A double buck (described in Section 5.9) may be needed.

2. Another error is not taking into account parasitic capacitance of the inductor windings and reverse current in the flywheel diode.

   These factors can be ignored in low to medium voltage DC applications, but not in applications where the supply voltage is more than about 200 V. The current peak through the MOSFET at turn-on can be high enough to trip the current sense circuit, resulting in erratic switching. An RC filter between the current sense resistor and the current sense input of the integrated circuit may be necessary. A 2.2 kΩ series resistor followed by a 100 pF shunt capacitor to ground should be sufficient.

3. Trying to operate a fixed frequency buck circuit (peak current mode control) with a duty cycle above 50% will result in instability.

   The switching frequency will jump to half the nominal value, known as subharmonic switching. However, there is a simple solution: if the load voltage is more than 50% of the supply voltage, constant off-time switching should be used. In the case of the HV9910, this is achieved by connecting the timing resistor to the Gate pin instead of to ground (0 V). If fixed frequency is required, or the controller IC does not have constant off-time mode, slope compensation is required. Slope compensation is described in Chapter 10 (switch power supplies).

Boost Converters

Chapter Outline

Boost converters are ideal for light-emitting diode (LED)–driver applications where the LED string voltage is greater than the input voltage. At the simplest, the applications can be low-voltage, battery-powered torches (flashlights) driving a short string of about LEDs. More complex applications can be backlights for liquid crystal display (LCD) screens, powered from 12–24 V, driving longer LED strings under demanding pulse width modulation (PWM) schemes and used in televisions or computer monitors.

6.1 Charge Pump Boost Converters

Low-power boost converters can use a charge pump instead of an inductor. A charge pump has many capacitors and switches, which are used to raise the output voltage above the supply voltage. This works by charging a capacitor from the supply voltage and then rearranging the switches so that this charge is moved to the next capacitor in line. This process repeats many times until the last capacitor in the chain is charged. A high-output voltage is achieved, which supplies current for the LED load. Fig. 6.1 shows a simple charge pump with 1.5 times boost ratio.

The two capacitors used as the charge pump are connected by single-pole changeover switches. The supply voltage first charges the capacitors in series, so each charges to half the input voltage. Then all four changeover switches operate so that the two capacitors are connected in parallel, one side connected to the input voltage rail and the other side connected to the output. Now the output voltage equals the input voltage plus half the input voltage (supplied from the capacitors). Thus we have a 1.5 times output/input voltage ratio.

The switch configuration can be changed so that the charge pump capacitors charge during alternate clock phases (180-degrees apart). While the first one is charging the second one is connected to the output and is discharging. The discharging capacitor connected so that its voltage adds to the input voltage, giving 2 times output/input voltage ratio. The switch

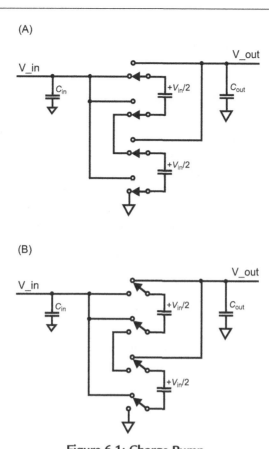

Figure 6.1: Charge Pump.
(A) Capacitors charge in series and (B) discharge in parallel.

configuration is changed rapidly, so that the first capacitor drives the output and the second capacitor is recharged, before the capacitor's stored charge has dropped significantly. This ensures a low-level output voltage ripple.

An example charge pump IC is the AP3606, from BCD Semiconductor. The output boost voltage is small, just 1.5 times the input voltage, but its purpose is to drive four white LEDs, with a forward voltage (V_f) of about 3.5 V, from a 2.7- to 4.2-V battery. The charge pump boosts the output voltage to a common anode rail. Each of the four LEDs has a cathode connected to a separate 20-mA current regulator. This type of charge pump is used in portable applications where a LCD display backlight needs to be driven.

Another charge pump IC is the LTC3215. This is intended for driving a single white LED at high current, up to 700 mA, with a 2.9- to 4.4-V input. The boost converter step-up ratio can be configured as 1×, 1.5×, or 2×. The target application is handheld mobile phones and PDAs, to provide a torch (flashlight) function or camera flash. Due to the high-current capability, a soft-start function is included in the IC. This prevents in-rush current, which could cause the battery voltage to drop significantly and reset the system.

6.2 Inductor-Based Boost Converters

Instead of using many small-energy transfers in a charge pump circuit, an inductor can be used to store and release energy quickly. Looking first at Fig. 6.2A, the basic process is to connect the inductor (*L*) across the power supply rails (V_IN to Ground) by closing switch S1 while S2 remains open. The current increases and the magnetic core stores energy. Then the ground connection is broken by opening S1, while at the same time closing S2. As there cannot be a sudden change in the inductor's current, the output capacitor is charged is charged by this current. The current will gradually reduce as the inductor's energy is transferred to the capacitor. The voltage across the output will depend on the capacitor value initially because the energy released by the inductor is now all stored in the capacitor.

Subsequent switching cycles of switches S1 and S2 will gradually increase the output voltage until there is a balance of average inductor current and load current. If the load is open circuit, the output voltage could rise very high. In a real circuit, to avoid damage to the circuit components, some form of overvoltage protection would be required.

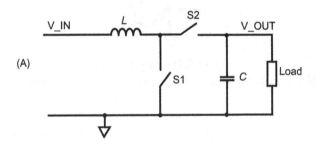

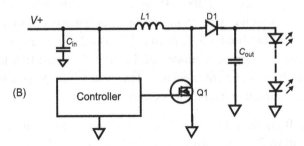

Figure 6.2: Simplified Inductor-Based Boost Converter Circuits.
(A) Theoretical implementation with switches and (B) practical implementation using MOSFET and diode.

A simplified practical inductor-based boost converter circuit is shown in Fig. 6.2B. Here switch S1 is replaced by MOSFET Q1 and is operated by a controller circuit (typically a PWM-controller IC). Switch S2 is a diode, D1.

If the boost converter is used in continuous conduction mode (more on this later), diode D1 is conducting in the forward direction when MOSFET Q1 turns on. So when the MOSFET conducts, the diode will conduct in the reverse direction until a depletion layer forms at the P–N junction. The speed at which the depletion layer forms is known as the reverse recovery time. To minimize switching losses, an ultrafast diode is required, with switching times much less than 75 ns.

If the boost converter is used in a discontinuous conduction mode, so that the inductor (and hence diode) current drops to zero before the MOSFET turns on, the diode can be a slower type without any switching loss penalty. However, it still needs to have a reverse recovery time less than, say, 250 ns to ensure that reverse current is minimized, as the period that the diode current is at zero could be as short.

We need some equations to work out the boost converter behavior. To start with, when the MOSFET Q1 turns on and connects the supply voltage across the inductor, we need to calculate how the current changes over time. The familiar equation for an inductor applies:

$$E = -L\, di/dt.$$

Where E is the electromagnetic force (EMF) or voltage, L is the inductance and di/dt is the rate of change in current (A/s). When a voltage (E) is applied across and inductor (L), the current rises at a constant rate (di/dt).

Other basic equations are for energy storage:

$$\text{Energy in inductor} = E = \frac{L}{2} \cdot I^2$$

$$\text{Energy in capacitor} = E = \frac{C}{2} \cdot V^2$$

So if, when the diode conducts, energy is totally transferred from the inductor to the capacitor:

$$\frac{L}{2} \cdot I^2 = \frac{C}{2} \cdot V^2$$

$$V = I \cdot \sqrt{\frac{L}{C}}$$

The voltage boost is proportional to the peak current in the inductor.

Normally, an inductor-based boost converter would only be used when the output voltage minimum is about 1.5 times the input voltage. If the output voltage is only a small amount above the supply voltage, there is the danger that if the output voltage falls (perhaps due to the LED temperature rising), or the input voltage rises (perhaps due to tolerances in the supply), the supply voltage could be greater than the load voltage. In this case, uncontrolled current could flow and damage the circuit.

At low-boost ratios, it is a good idea to connect a diode from input to output. This provides a path for the initial output capacitor charging, to avoid overcurrent in the inductor.

- The inductor-based converter can easily be designed to operate at efficiencies greater than 90%.
- Both the MOSFET and LED string can be connected to a common ground. This simplifies sensing of the LED current, unlike the buck converter where we have to choose either a high-side MOSFET driver or a high-side current sensor.
- The input current can be continuous or discontinuous. Continuous mode can only be used if the output is no more than 6 times the input voltage. Why?
  - The maximum duty cycle of operation when the output is 6 times the input can be computed as
    $$D_{max} = 1 - \frac{\eta_{min} \cdot V_{in\,min}}{V_{o\,max}}$$, so if $\eta_{min} = 0.9$, $D_{max} = 0.85$, which is the limit for most
    controllers. This allows enough off-time for the energy in the inductor to be transferred to the output capacitor and load.
  - If continuous conduction is used, it is easier to filter the input ripple current and thus easier to meet any required conducted electromagnetic interference (EMI) standards.
- Discontinuous mode is always used when the output voltage is more than 6 times higher than the input voltage, but can be used at lower voltage ratios if required. Equalization of a discontinuous mode circuit is simpler than for a continuous mode circuit because first-order equalization (a simple capacitor) can be used, and is sometimes favored to provide low-boost ratios for that reason.

6.2.1 Low-Voltage Inductive Boost Converters

Let us first look at boost converters operating from a low voltage. These are used instead of charge pumps when the boost ratio is higher than what a charge pump can supply. They are often used in portable applications, such as backlighting and LCD display in laptop computers, PDAs, handheld terminals, etc.

An example of a low-voltage boost controller IC for LED driving is ON-Semiconductor's NCP5050. This is optimized for driving two white LEDs at 200 mA from a 2.7- to 5.5-V supply. Higher currents are possible, but this depends on the number of LEDs and the input supply voltage: graphs given in the datasheet should be studied carefully to determine the current limit for the supply that one has. Obvious applications for this IC are camera flash or torch (flashlight), where the supply is a nominal 3.2-V Li-ion battery or two 1.5-V alkaline cells in series. The limitation of this IC is the current accuracy; even with 1% tolerance current sense resistors, the output current tolerance is ±20%.

The Intersil ISL97634 is a low-voltage boost LED driver, intended for backlighting LCD displays in cell phones and satellite navigation units. It can operate from 2.4- to 5.5-V input and output up to 26 V. The internal boost switch is rated at 400 mA, allowing a 30-mA LED current at 26 V, but up to 70 mA with just a 14-V load. This device features a high-frequency PWM dimming capability, up to 32 kHz, thus avoiding the audible frequency range. Fast dimming is achieved by disconnecting the LEDs. This is achieved by a second internal switch, connected in series with the LED string.

Another example of a low-voltage boost controller is the TPS61040 from Texas Instruments (TI). This operates from a 1.8- to 6-V supply and can be configured for constant voltage or constant current output. For constant voltage output, the feedback pin is connected to a potential divider placed across the output. But to get constant current output suitable for driving LEDs, the feedback pin is connected to a current sense resistor placed between the LED cathode and ground (0 V). This IC is for lower power applications, as the internal switch is rated at 400 mA, and the datasheet example shows the boost circuit driving four white LEDs at 15 mA from a 2.7- to 6-V supply. However, this IC is very small and only needs six external components, so the complete circuit can fit into a small space and is ideal for some portable applications. A typical circuit driving four white LEDs with 10 mA is shown in Fig. 6.3.

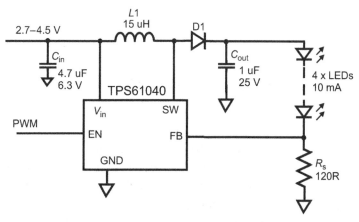

Figure 6.3: Low-Voltage Inductive Boost Circuit.
PWM, Pulse width modulation.

An optional Zener diode could be connected across the LEDs, to protect the driver circuit in case of an open LED. However, in a sealed module, replacement of the LEDs is almost impossible. Most applications would have no open LED protection because, in the case of an open LED, the whole backlight circuit would be replaced and so the damage to the driver circuit would be immaterial.

6.2.2 Medium-Voltage Inductive Boost Converters

Microchip's HV9912 integrated circuit is a closed-loop, peak current–controlled, switch-mode converter LED driver. It can operate over a 10–90 V input voltage range, so is suitable for operating from standard 12-, 24-, or 48-V supplies. It has been used for television LCD screen backlights, using a 24-V supply and driving a long LED string at 350 mA with up to 80-V output.

The HV9912 has built-in features to overcome the disadvantages of the basic boost converter. In particular, it features a disconnect MOSFET driver output. The external MOSFET driven from this output can be used to disconnect the LED strings during short circuit, or input overvoltage, conditions. This disconnect MOSFET is also used by the HV9912 to dramatically improve the PWM dimming response of the converter (more details at the end of this Section), by disconnecting the LED string when the PWM dimming control input is at Logic 0. This disconnection prevents the output capacitor from discharging through the LEDs and gives a short turn-on/turn-off periods.

Linear Technology's LTC3783 has similar functionality to the HV9912, although this part operates from a lower-voltage supply (6–16 V input). It is not a drop-in replacement and overvoltage protection (OVP) and oscillator-timing functions require a small-design change. For example, in the LTC3783 the OVP threshold is 1.23 V, so this is much lower than the 5-V threshold in the HV9912 and feedback resistors will have to be changed. The LTC3783 oscillator frequency is set by:

$$F_{osc} = 6 \times 10^9 / R_T, \text{ where } R_T \text{ is the timing resistor. So for a 20-kHz operation, } R_T = 300 \text{ k}\Omega.$$

The LTC3783 feedback transconductance amplifier gain is 588 μA/V, which is almost the same as in the HV9912, so compensation components can be the same values with no significant effect.

Monolithic power (MPS) did have the MP4012, which was virtually the same design as the HV9912. However, this part has now been superseded by the MP4013B, which has a maximum input voltage of 26 V. The switching current limit reference voltage is fixed at 485 mV and the output current sense reference voltage is 600 mV (the HV9912 uses external resistors to set the reference voltages at the C_LIM pin and the I_REF pin). The feedback amplifier has a transconductance gain of $G_m = 370$ μA/V, which will affect the compensation

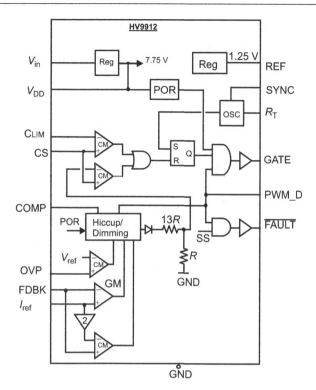

Figure 6.4: HV9912 Internal Structure (Simplified).
OVP, Overvoltage protection.

component selection. The OVP reference voltage is 5 V, like the HV9912. So later function descriptions and design examples using the HV9912 also apply to the MP4013B, provided that the reference voltages described about are substituted into the equations as necessary.

The most significant functions within the HV9912 are shown in Fig. 6.4.

The internal high-voltage regulator in the HV9912 provides a regulated 7.75-V V_{DD} from a 10–90 V input, which is used to power the IC. This voltage range is good for most boost applications, but the IC can also be used in buck and SEPIC circuits when accurate current control is required. In a high-voltage buck application, a Zener diode could be added in series with the input to allow an even higher operating voltage, or to reduce the power dissipated by the IC.

The V_{DD} pin of the IC can be overdriven (if necessary) with an external voltage source fed through a low voltage (>10 V), low-current diode. The diode will help to prevent damage to the HV9912 if the external voltage becomes less than the internally regulated voltage. The maximum steady state voltage that can be applied to the HV9912 V_{DD} pin is 12 V (with a transient voltage rating of 13.5 V). Allowing for the diode forward voltage drop a 12 V ± 5% power supply would be ideal.

The HV9912 includes a buffered 1.25-V, 2% accurate reference voltage. This reference voltage can be used to set the current reference level, as well as the input current limit level, by connecting potential divider networks between the REF pin and the I_{REF} and CLIM pins. This reference is also used internally to set the overvoltage set point.

Using an external resistor, we can set the oscillator timing of the HV9912. If the resistor is connected between the R_T and GND pins, the converter operates in a constant frequency mode, whereas if it is connected between the R_T and GATE pins, the converter operates in a constant off-time mode (slope compensation is not necessary to stabilize the converter operating in a constant off-time). In both cases, the clock period or off-time can be set to any value between 2.8 μs and 40 μs using the oscillator timing equation given in Section 6.4.12.

Multiple HV9912 ICs can be synchronized to a single-switching frequency by connecting the SYNC pins of all the IC together. This is sometimes necessary in RGB lighting systems, or when EMI filters are designed to remove a certain frequency.

Closed-loop control is achieved by connecting the output current sense signal to the FDBK pin and the current reference signal to the I_{REF} pin. The HV9912 tries to keep the feedback signal equal to the voltage on the I_{REF} pin. If the feedback is too high, indicating that the current is above the required level, the MOSFET switching is stopped. When the feedback falls below the voltage at the I_{REF} pin, switching is started again. The transconductance feedback amplifier connected to the FDBK pin and the I_{REF} pin, has a gain of 550 μA/V. This gain is used in equation to calculate the value of compensation components so, if other driver ICs are used, their feedback gain must be used instead in these calculations.

The compensation network is connected to the COMP pin (output of the transconductance op-amp). What is not shown in Fig. 6.4 is that the output of the amplifier has a switch controlled by the PWM dimming signal. When the PWM dimming signal is low, this switch disconnects the output of the amplifier. Thus, the capacitor(s) in the compensation network hold the voltage while the PWM signal is low. When the PWM dimming signal goes high again, the compensation network is reconnected to the amplifier. This ensures that the converter starts at the correct operating point and a very good PWM dimming response is obtained without having to design a fast controller.

The $\overline{\text{FAULT}}$ pin is used to drive an external disconnect MOSFET (Fig. 6.5). During the start-up of the HV9912, the $\overline{\text{FAULT}}$ pin is held low and once the IC starts-up the pin is pulled high. This connects the LEDs in the circuit and the boost converter powers up the LEDs. In case of an output overvoltage condition or an output short circuit condition, the $\overline{\text{FAULT}}$ pin is pulled low and an external MOSFET switched off to disconnect the LEDs.

The $\overline{\text{FAULT}}$ pin is also controlled by the PWM dimming signal, so that the pin is high when the PWM dimming signal is high and vice versa. This disconnects the LEDs and makes sure that the output capacitor does not have to be charged/discharged every PWM dimming cycle. The

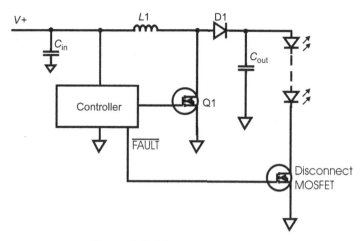

Figure 6.5: Disconnect MOSFET.

PWM dimming input to the $\overline{\text{FAULT}}$ pin and the output of the protection circuitry are logically AND'ed to make sure that the protection circuit overrides the PWM input to the $\overline{\text{FAULT}}$ pin.

Output short circuit protection is provided by a comparator that triggers when the output current sense voltage (at the FDBK pin) is twice that of the reference voltage (at the I_{REF} pin). The output overvoltage protection is activated when the voltage at the OVP pin exceeds 5 V. A fault condition will exist when either of these signals are active and will result in both the GATE pin and the $\overline{\text{FAULT}}$ pin being pulled low. Once the IC goes into the fault mode, the capacitor on the COMP pin is used for timing until a reset occurs. At the end of the timing period, the HV9912 will attempt to restart. If the fault condition persists, the HV9912 will continually attempt restarts until power is removed or the fault clears.

Linear dimming is achieved by varying the voltage level at the I_{REF} pin. This can be done either with a potentiometer from the REF pin or from an external voltage source and a resistor divider. This allows the current to be linearly dimmed. However, note that once the voltage at the I_{REF} pin is lowered to a very small value, the offsets of the output short circuit current comparator might cause the HV9912 fault mode to activate improperly. The power to the IC will have to be recycled to start-up the circuit again. To prevent this false triggering, it is advisable to limit the minimum voltage at the I_{REF} pin to about 20–30 mV.

The features included in the HV9912 help achieve a very fast PWM dimming response in spite of the shortcomings of the boost converter. The PWM dimming signal controls three nodes in the IC.

- Gate signal to the switching MOSFET.
- Gate signal to the disconnect MOSFET.
- Output connection of the transconductance op-amp.

When PWMD is high, the gates of both the switching MOSFET and the disconnect MOSFET are enabled. At the same time, the output of the transconductance op-amp is connected to the compensation network. This allows the boost converter to operate normally.

When PWMD goes low, the GATE of the switching MOSFET is disabled to stop energy transfer from the input to the output. However, this does not prevent the output capacitor from discharging into the LEDs, causing a large-decay time for the LED current. This discharge of the capacitor also means that when the circuit restarts, the output capacitor has to charge again, causing an increase in the rise time of the LED current. This problem becomes more prominent with larger-output capacitors. Thus, it is important to prevent the discharge of the output capacitor. This is done by turning off the disconnect MOSFET. This causes the LED current to fall to zero almost instantaneously. As the output capacitor does not discharge, there is no necessity to charge the capacitor when PWMD goes high. This enables a very fast rise time as well.

So what happens if our controller does not have a sample-and-hold switch on the output of the feedback amplifier? When PWMD goes low, the output current goes to zero. This means that the feedback amplifier sees a very large-error signal across its input terminals, which would cause the voltage across the compensation capacitor to increase to the positive rail. Thus, when the PWMD signal goes high again, the large voltage across the compensation network, which dictates the peak inductor current value, will cause a large spike in the LED current. The current will come back into regulation eventually, but how long this takes will depend on the speed of the controller.

The HV9912 disconnects the output of the amplifier from the compensation network when PWMD goes low, which helps to keep the voltage at the compensation unchanged. Thus, when PWMD goes high again, the circuit will already be at the steady state condition, eliminating the large turn-on spike in the LED current.

6.2.3 Multichannel Boost Converters

Some manufacturers have integrated multiple (typically 3) boost converters into a single package. This has been done typically for LCD backlights where one channel is assigned to drive each of red, green, and blue LEDs. Some manufacturers have made four-channel devices; this is because the green color LEDs have a relatively low-light emission, so the channels are assigned red, green, and blue. As green has two channels, the brightness levels are now fairly equal.

To reduce EMI, the channels are usually phased, so that the clock signals are 120-degrees apart in a 3-channel (3-phase) system or 90-degrees apart in a 4-channel (4-phase) system. By phasing there is no large-input current spike as all channels switch at the same time. This would require large-input capacitors for decoupling, to provide this current. In a phased system, the stresses on the input capacitors are reduced and smaller capacitors can be used.

Apart from the phased clocks, each of the boost converters can be considered as a single stage and normal design rules apply.

6.3 Boost Converter Operating Modes

An inductive boost converter can be operated in two modes: either continuous conduction mode (CCM) or discontinuous conduction mode (DCM). The mode of operation of the boost converter is determined by the waveform of the inductor current. Fig. 6.6A is the inductor current waveform for a CCM boost converter, whereas Fig. 6.6B is the inductor current waveform for a DCM boost converter.

The CCM boost converter is used when the maximum step-up ratio (ratio of output voltage to input voltage) is less than or equal to 6. If larger-boost ratios are required, the DCM boost converter is used. However, in DCM, the inductor current has large-peak values, which increases the core losses in the inductor. Thus DCM boost converters are typically less efficient than CCM boost converters, can create more EMI problems, and are usually limited to lower-power levels.

Inductive boost converters have some disadvantages, especially when used as LED drivers, due to the low-dynamic impedance of the LED string.

- The output current of the boost converter is a pulsed waveform. Thus, a large-output capacitor is required to reduce the ripple in the LED current.
- The large-output capacitor makes PWM dimming more challenging. Turning the boost converter on and off to achieve PWM dimming means the capacitor will have to be

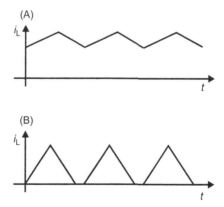

Figure 6.6: Inductor current (A) continuous conduction mode (CCM) and (B) discontinuous conduction mode (DCM).

charged and discharged every PWM dimming cycle. This increases the rise and fall times of the LED current.
- Open loop control of the boost converter to control the LED current (as in the case of a HV9910-based buck control) is not possible. Closed-loop control is required to stabilize the converter. This also complicates PWM dimming, as the controller will have to have a large bandwidth to achieve the required response times.
- There is no control over the output current during output short circuit conditions. There is a path from the input to the output via a diode and inductor, so turning off the switching MOSFET will have no effect on the short circuit current.
- There will be a surge of current into the LEDs if an input voltage transient raises the input voltage above the LED string voltage. If the surge current is high enough, the LEDs will be damaged.

6.4 Design of a Continuous Conduction Mode Boost Circuit

As a reminder, CCM is valid when the output voltage is between 1.5 and 6 times the input voltage.

6.4.1 Design Specification

Input voltage range = 22–26 V

LED string voltage range = 40–70 V

LED current = 350 mA

LED current ripple = 10% (35 mA)

LED string dynamic impedance = 18 Ω

Desired efficiency > 90%

6.4.2 Typical Circuit

A typical boost converter circuit is shown in Fig. 6.7.

6.4.3 Selecting the Switching Frequency (f$_s$)

For low-voltage applications (output voltage < 100 V), and moderate-power levels (<30 W), a switching frequency of f_s = 200 kHz is a good compromise between switching power loss and size of the components. At higher voltage or power levels, the switching frequency might have to be reduced to lower the switching losses in the external MOSFET.

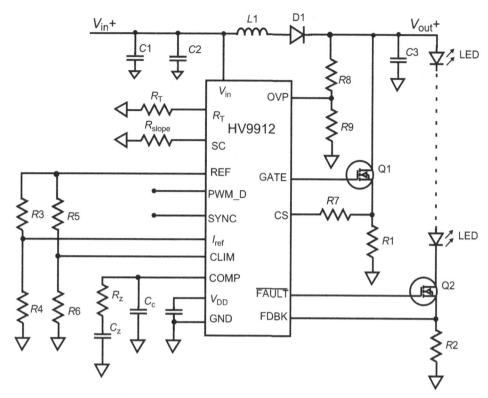

Figure 6.7: Continuous Mode Boost Converter.

6.4.4 Computing the Maximum Duty Cycle (D_{max})

The maximum duty cycle of operation can be computed as

$$D_{max} = 1 - \frac{\eta_{min} \cdot V_{in\,min}}{V_{o\,max}}$$

$$= 0.717$$

Note: If $D_{max} > 0.85$, the step-up ratio is too large. The converter cannot operate in CCM and has to be operated in DCM to achieve the required step-up ratio.

6.4.5 Computing the Maximum Inductor Current ($I_{in\,max}$)

The maximum input current is

$$I_{in\,max} = \frac{V_{o\,max} \cdot I_{o\,max}}{\eta_{min} \cdot V_{in\,max}}$$

$$= 1.24\,A$$

6.4.6 Computing the Input Inductor Value (L1)

The input inductor can be computed by assuming a 25% peak-to-peak ripple in the inductor current at minimum input voltage.

$$L1 = \frac{V_{\text{in max}} \cdot D_{\text{max}}}{0.25 \cdot I_{\text{in max}} \cdot f_s}$$

$$= 254 \; \mu\text{H}$$

Choose a standard 330-μH inductor. To achieve 90% efficiency at the minimum input voltage, the power loss in the inductor has to be limited to around 2–3% of the total output power. Using a 3% loss in the inductor

$$P_{\text{ind}} = 0.03 \cdot V_{o\,\text{max}} \cdot I_{o\,\text{max}}$$

$$= 0.735 \, \text{W}$$

Assuming a 80–20% split in the inductor losses between resistive and core losses, the DC resistance (DCR) of the chosen inductor has to be less than

$$\text{DCR} < \frac{0.8 \cdot P_{\text{ind}}}{I_{\text{in max}}^2}$$

$$\Rightarrow \text{DCR} < 0.38 \, \Omega$$

The saturation current of the inductor has to be at least 20% higher than its peak current; otherwise the core losses will be too great.

$$I_{\text{sat}} = 1.2 \cdot I_{\text{in max}} \cdot \left(1 + \frac{0.25}{2}\right)$$

$$= 1.7 \, \text{A}$$

Thus $L1$ is a 330-μH inductor with a DCR about 0.38 Ω and a saturation current greater than 1.7 A.

Note: Choosing an inductor with an RMS current rating equal to $I_{\text{in max}}$ would also yield acceptable results, although meeting the minimum efficiency requirement might not be possible.

6.4.7 Choosing the Switching MOSFET (Q1)

The maximum voltage across the MOSFET in a boost converter is equal to the output voltage. Using a 20% overhead to account to switching spikes, the minimum voltage rating of the MOSFET has to be

$$V_{FET} = 1.2 \cdot V_{omax}$$
$$= 84 \text{ V}$$

The RMS current through the MOSFET is

$$I_{FET} \approx I_{in\,max} \cdot \sqrt{D_{max}}$$
$$= 1.05 \text{ A}$$

To get the best performance from the converter, the MOSFET chosen has to have a current rating about 3 times the MOSFET RMS current with minimum gate charge Q_g. The higher current rating gives low-conduction losses, even at high silicon junction temperatures (resistance increases with temperature). It is recommended that for designs with the HV9912, the gate charge of the chosen MOSFET be less than 25 nC.

The switching device chosen for this application is a 100 V, 4.5 A MOSFET with a Q_g of 11 nC.

6.4.8 Choosing the Switching Diode (D1)

The voltage rating of the diode is the same as the voltage rating of the MOSFET (100 V). The average current through the diode is equal to the maximum output current (350 mA). Although the average current through the diode is only 350 mA, the diode carries the full-input current $I_{in\,max}$ for short durations of time. Thus, it is a better design approach to choose the current rating of the diode somewhere in between the maximum input current and the average output current (preferably closer to the maximum input current). Thus, for this design, the diode chosen is a 100 V, 1 A Schottky diode.

6.4.9 Choosing the Output Capacitor (C$_o$)

The value of the output capacitor C_o (labeled C3 in Fig. 5.8) depends on the dynamic resistance of the LED, the ripple current desired in the LED string and the LED current. In designs using the HV9912, a larger-output capacitor (lower-output current ripple) will yield better PWM dimming results. The capacitor required to filter the current appropriately will be designed by considering the fundamental component of the diode current only.

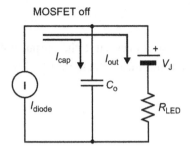

MOSFET off

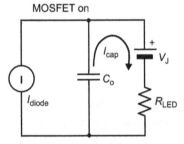

MOSFET on

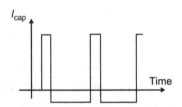

Figure 6.8: Model of Boost Converter Output.

Figure 6.9: Charge and Discharge Cycle of Output Capacitor.

The output stage of the boost converter is modeled in Fig. 6.8, where the LEDs are modeled as a constant voltage load with series dynamic impedance.

The output impedance (parallel combination of R_{LED} and C_o) is driven by the diode current. The waveform of the capacitor current in steady state is shown in Fig. 6.9; the capacitor is charged during the off-time, as the energy stored in the inductor is transferred to the capacitor. Whilst the MOSFET is turned on and energy is being stored in the inductor, the capacitor is discharged by the load.

Using the 10% peak-to-peak current ripple given in the design parameters table, the maximum voltage ripple across the LED string has to be

$$\Delta v_{p-p} = \Delta I_o \cdot R_{LED}$$
$$= 0.63\,\text{V}$$

Assuming a constant discharging current of 350 mA when the switch is ON, the equation for the voltage across the capacitor can be written as

$$I_{o\max} = C_o \cdot \frac{\Delta v_{p-p}}{D_{\max} \cdot T_s}$$

Substituting values into this equation, we can calculate the value for C_o.

$$C_o = \frac{I_{o\max} \cdot D_{\max}}{\Delta v_{p\text{-}p} \cdot f_s}$$
$$= 1.99 \, \mu F$$

The RMS current through the capacitor can be given by

$$I_{rms} = \sqrt{D_{\max} \cdot I_{o\max}{}^2 + (1 - D_{\max}) \cdot (I_{in\max} - I_{o\max})^2}$$
$$= 0.56 \, A$$

In this case, a parallel combination of two 1-μF, 100-V metal polypropylene capacitors is chosen.

Note: The proper types of capacitors to use are either metal film capacitors or ceramic capacitors, as they are capable of carrying this high-ripple current. Although ceramic capacitors are smaller in size and capable of carrying the ripple current, they cause a lot of audible noise during PWM dimming as they have a piezoelectric effect. Also, high-value ceramic capacitors are normally only rated up to 50 V. Thus metal polypropylene (or any other metal film) capacitors are the ideal choice for LED drivers if PWM dimming is required.

6.4.10 Choosing the Disconnect MOSFET (Q2)

The disconnect MOSFET should have the same voltage rating as the switching MOSFET Q1. The on-state resistance of the MOSFET at room temperature ($R_{on,25°C}$) has to be chosen based on a 1% power loss in Q2 at full-load current. Thus,

$$R_{on,25°C} = \frac{0.01 \cdot V_{o\max}}{I_{o\max} \cdot 1.4}$$
$$= 1.43 \, \Omega$$

The 1.4 multiplication factor is included to account for the increase in the on-resistance due to a rise in junction temperature. In this case, a MOSFET with high-gate charge, Q_g, can be chosen if desired (as it is not switching regularly). A high Q_g MOSFET will slow down the turn-on and turn-off times. In this case, the MOSFET chosen is a 100 V, 0.7 Ω, SOT-89 MOSFET with a Q_g of 5 nC.

6.4.11 Choosing the Input Capacitors (C1 and C2)

The values of input capacitors *C1* and *C2* have to be calculated to meet closed-loop stability requirements. The connection from the power source to the boost converter circuit will have some resistance, R_{source}, and some inductance, L_{source}. These feed across the input capacitor ($C1$ and $C2$) and so form an LC resonant circuit. To prevent interference with the control loop, the resonant frequency should be arranged to be 40% or less of the switching frequency.

How do we determine the inductance L_{source}? A pair of 22AWG connecting wires 1-ft. (30-cm) long will have an inductance of about 1 μH. This is a good starting point. If necessary, the wires can be twisted together to reduce the inductance.

With a 200-kHz switching frequency, the resonant frequency should be less than 80 kHz.

$$C_{in} \geq \frac{1}{(2 \cdot \pi \cdot f_{LC})^2 \cdot L_{source}} = 3.95 \, \mu F$$

$C1 = C2 = 2.2$ μF, 50 V ceramic.

The magnitude of the reflected converter impedance at the LC resonant frequency is given by:

$$R_{eq} = (1 - D_{max})^2 \cdot R_{LED}$$
$$R_{eq} = (1 - 0.717)^2 \cdot 18$$
$$R_{eq} = 1.4416 \Omega$$

$$R_{source,max} = 1.44 \Omega$$

6.4.12 Choosing the Timing Resistor (R_T)

The HV9912 oscillator has an 18-pF capacitor charged by a current mirror circuit. An external timing resistor R_T provides a reference current for the current mirror. When R_T is connected to 0 V, current flows and the timing process begins. When charged to a certain voltage, the R_S flip-flop is set, the capacitor is discharged, and the timing process starts again. The timing resistor value can be calculated by using the equation:

$$\frac{1}{f_s} \approx R_T \cdot 18 \, pF$$

In this case, for a constant 200-kHz switching frequency, the timing resistor value works out to approximately 270 kΩ. This resistor needs to be connected between the R_T pin and GND as shown in the typical circuit.

If using a different controller IC, the equation may have to be modified. For example the HV9911 has a timing capacitor of 11 pF, rather than the 18 pF of the HV9912, so the timing resistor would have to be proportionally larger; that is, $18/11 \times 270$ kΩ $= 442$ kΩ.

6.4.13 Choosing the Two Current Sense Resistors (R1 and R2)

The value of output current sense resistor $R2$ (R_{cs} in later calculations) is calculated to limit its power dissipation to about 0.15 W, so that a 1/4 W resistor can be used. Using this criterion,

$$R2 = \frac{0.15\,\text{W}}{I_{o\,\text{max}}^{2}}$$
$$= 1.22\,\Omega$$

In this case, the resistor chosen is a 1.24 Ω, 1/4 W, 1% resistor.

The MOSFET current sense resistor $R1$ (R_s in later calculations) is calculated by limiting the voltage across the resistor to about 250 mV at maximum input current. A 12.5% margin is added to the maximum input current to allow for inductor and other component tolerances.

$$R1 = \frac{0.25}{1.125 \cdot I_{in\,\text{max}}}$$
$$= 0.18\,\Omega$$

The power dissipated in this resistor is

$$P_{R1} = I_{\text{FET}}^{2} \cdot R1$$
$$= 0.2\,\text{W}$$

Thus, the chosen current sense resistor is a 0.18 Ω, 0.25 W, 1% resistor.

6.4.14 Selecting the Current Reference Resistors (R3 and R4)

The voltage at the current reference pin I_{REF} can be set either by using the reference voltage provided at the REF pin (through a voltage divider) or with an external voltage source. In the present design, it is assumed that the voltage at the I_{REF} pin is set using a voltage divider from the REF pin. The current reference resistors $R3$ and $R4$ can be computed using the following two equations:

$$R3 + R4 = \frac{1.25\,\text{V}}{50\,\mu\text{A}} = 25\,\text{k}\Omega$$

$$\frac{1.25\,\text{V}}{R3 + R4} \cdot R4 = I_{o\,\text{max}} \cdot R2$$

For this design, the values of the two resistors can be computed to be

$$R_{R2} = 8.68\,k\Omega,\ 1/8\,W,\ 1\%$$
$$R_{R1} = 16.32\,k\Omega,\ 1/8\,W,\ 1\%$$

6.4.15 *Programming the Slope Compensation (R_{slope} and R7)*

As the boost inductor being designed is operating at constant frequency, slope compensation is required to ensure the stability of the converter. The slope added to the current sense signal has to be one-half the maximum down slope of the inductor current to ensure stability of the peak current mode control scheme for all operating conditions. This can easily be achieved by the proper selection of the two slope compensation resistors R_{slope} and $R7$.

For the present design, the down slope of the inductor current is

$$DS = \frac{V_{o\,max} - V_{in\,min}}{L}$$
$$= 0.145\ A/\mu s$$

The programming resistors can then be calculated as

$$R_{slope} = \frac{10 \cdot R7 \cdot f_s}{DS(A/\mu s) \cdot 10^6 \cdot R1}$$

Assuming that $R7 = 1\ k\Omega$,

$$R_{slope} = \frac{10 \cdot 1k \cdot 200k}{0.2682 \cdot 10^6 \cdot 0.15}$$
$$= 76.62\,k\Omega$$

Note: The maximum current that can be sourced out of the SC pin is limited to 100 μA. This limits the minimum value of the R_{slope} resistor to 25 kΩ. If the equation for slope compensation produces a value R_{slope} less than this value, then $R7$ would have to be increased accordingly. It is recommended that R_{slope} be chosen in the range of 25–50 kΩ.

Based on this recommendation, the calculated values can be scaled by 0.51. The selected resistor values are

$$R7 = 510,\ 1/8\,W,\ 1\%$$
$$R_{slope} = 39k,\ 1/8\,W,\ 1\%$$

6.4.16 Setting the Inductor Current Limit (R5 and R6)

The inductor current limit value depends on two factors, the maximum inductor current and the slope compensation signal added to the sensed current. Another resistor divider, connected to the REF pin, sets this current limit. The voltage at the CLIM pin can be computed as

$$V_{\text{CLIM}} \geq 1.35 \cdot I_{\text{in max}} \cdot R1 + \frac{4.5 \cdot R7}{R_{\text{slope}}}$$

This equation assumes that the current limit level is set at about 120% of the maximum inductor current $I_{\text{in max}}$ and that the operating duty cycle is at 90% (maximum for the HV9912).

For this design,

$$\begin{aligned} V_{\text{CLIM}} &= 1.35 \cdot 1.24 \cdot 0.18 + \frac{4.5 \cdot 510}{39k} \\ &= 0.36\,\text{V} \end{aligned}$$

We need a potential divider to give 0.36 V from a 1.25 V reference. Using a maximum current sourced out of REF pin of 50 µA the two resistors in series should be >25 kΩ, and can be calculated as:

$$R5 = 20k, 1/8\,\text{W}, 1\%$$
$$R6 = 8.06k, 1/8\,\text{W}, 1\%$$

Note: It is recommended that no capacitor be connected at the CLIM pin because this would affect start-up conditions while it is charging.

6.4.17 Capacitors at V_DD and REF pins

It is recommended that bypass capacitors be connected to both V_{DD} and REF pins. For the V_{DD} pin, the capacitor recommended is a 1-µF ceramic chip capacitor. If the design uses switching MOSFETs that have a high-gate charge ($Q_g > 15$ nC), the capacitor value at the V_{DD} pin should be increased to 2.2 µF. Ceramic capacitors have a low impedance, so they are good for supplying a burst of gate drive current.

For the REF pin, the capacitor used is a 0.1-µF ceramic chip capacitor.

6.4.18 Setting the Overvoltage Trip Point (R8 and R9)

The overvoltage trip point can be set at a voltage 15%, or more, above the maximum steady state voltage. Using a 20% margin, the maximum output voltage during open LED condition will be

$$V_{open} = 1.2 \cdot V_{omax}$$
$$= 84 \, V$$

In the HV9912, the overvoltage comparator input has a 5-V reference. So, the resistors that set the overvoltage set point can be computed as:

$$R8 = \frac{(V_{open} - 5 \, V)^2}{0.1}$$
$$= 64 \, k\Omega$$

This equation will allow us to select a 0.125-W resistor by limiting the power dissipation in the resistor to 0.1 W.

$$R9 = \frac{R8}{(V_{open} - 5 \, V)} \cdot 5 \, V$$
$$= 3.95 \, k\Omega$$

The closest 1% resistor values are

$$R8 = 68k, \, 1/8 \, W, \, 1\%$$
$$R9 = 3.9k, \, 1/8 \, W, \, 1\%$$

If in a different controller IC, the overvoltage comparator reference voltage is not 5 V, substitute the new value in place of the 5 V used in both previous equations.

Note: The actual overvoltage point will vary from the desired point by ±5% due to the variation in the reference (see datasheet). For this design, it varies from 87.57 to 96.79 V.

6.4.19 Designing the Compensation Network

The compensation needed to stabilize the converter could be either a type-I circuit (a simple integrator) or a type-II circuit (an integrator with an additional pole-zero pair). The type of the compensation circuit required will be dependent on the phase of the power stage at the crossover frequency.

The loop gain of the closed-loop system is given by

$$\text{Loop gain} = R_s \cdot G_m \cdot Z_c(s) \cdot \frac{1}{15} \cdot \frac{1}{R_{cs}} \cdot G_{ps}(s)$$

Where G_m is the transconductance of the HV9912 internal op-amp (550 µA/V), $Z_c(s)$ is the impedance of the compensation network, and $G_{ps}(s)$ is the transfer function of the power stage. Please note that although the resistors give a 1:14 ratio, the overall effect when including the diode drop is effectively 1:15.

For the CCM boost converter in peak current control mode and for frequencies less than 1/10th of the switching frequency, the power stage transfer function is given by

$$G_{ps}(s) = \frac{(1-D_{max})}{2} \cdot \frac{1 - s \cdot \dfrac{L1}{(1-D_{max})^2 \cdot R_{LED}}}{1 + s \cdot \dfrac{R_{LED} \cdot C_o}{2}}$$

For the present design, choose a crossover frequency $0.01 \times f_s$, $f_c = 2$ kHz. The low crossover frequency will result in large values for C_c and C_z, which will indirectly provide a soft-start for the circuit. As the HV9912 does not depend on the speed of the controller circuit for the PWM dimming response, the low crossover frequency will not have an adverse effect on the PWM dimming rise and fall times.

$$G_{ps}(s) = \frac{0.283}{2} \cdot \frac{1 - s \cdot \dfrac{330 \cdot 10^{-6}}{(0.283)^2 \cdot 18}}{1 + s \cdot \dfrac{18 \cdot 2 \cdot 10^{-6}}{2}}$$

$$G_{ps}(s) = 0.1415 \cdot \frac{1 - s \cdot 2.28912 \cdot 10^{-4}}{1 + s \cdot 1.8 \cdot 10^{-5}}$$

Substituting $s = i \cdot (2\pi \cdot f_c)$, where $f_c = 2$ kHz, $s = i \cdot 12{,}566$.

$$G_{ps}(s) = 0.1415 \cdot \frac{1 - i \cdot 2.8766}{1 + i \cdot 0.226188}$$

At this frequency, the magnitude and frequency of the power stage transfer function [obtained by substituting $s = i \cdot (2\pi \cdot f_c)$ in the previous equation] are

$$\left| G_{ps}(s) \right|_{f_c = 2\,\text{kHz}} = A_{ps} = 0.40996$$

$$\angle G_{ps}(s) \big|_{f_c = 2\,\text{kHz}} = \phi_{ps} = -83.57°$$

To get a phase margin of about $\phi_m = 45$ degrees (the recommended phase margin range is 45–60 degrees), the phase boost required will be

$$\phi_{boost} = \phi_m - \phi_{ps} - 90°$$
$$= 38.57°$$

Based on the value of the phase boost required, the type of compensation can be determined.

$$\phi_{boost} \leq 0° \Rightarrow \text{Type-I controller}$$
$$0° \leq \phi_{boost} \leq 90° \Rightarrow \text{Type-II controller}$$
$$90° \leq \phi_{boost} \leq 180° \Rightarrow \text{Type-III controller}$$

As we need a 38.57°-phase boost, we need a type-II controller. Note that type-III controllers are usually not required to compensate a HV9912-based boost LED driver and thus will not be discussed further.

The implementations for the type-I and type-II systems for use with the HV9912 are given in Table 6.1.

The present CCM design needs a type-II controller. The equations needed to design the type-II controller are as follows:

$$K = \tan(45° + \frac{\phi_{boost}}{2})$$
$$= 2.077$$

Table 6.1: Compensation networks

Type	Circuit Diagram	Transfer Function
I	COMP — C_c	$Z_c(s) = \dfrac{1}{sC_c}$
II	COMP, C_z, R_z, C_c	$Z_c(s) = \dfrac{1}{s(C_c + C_z)} \cdot \dfrac{1 + s \cdot R_z \cdot C_z}{1 + s \cdot \dfrac{C_z \cdot C_c}{C_z + C_c} \cdot R_z}$

$$\omega_z = \frac{1}{R_z \cdot C_z} = \frac{2 \cdot \pi \cdot f_c}{K}$$
$$= 6{,}050 \ \text{rad/s}$$

$$\omega_p = \frac{C_z + C_p}{C_z \cdot C_p \cdot R_z} = (2 \cdot \pi \cdot f_c) \cdot K$$
$$= 26{,}100 \ \text{rad/s}$$

One more equation can be obtained by equating the magnitude of the loop gain to 1 at the crossover frequency.

$$R_s \cdot G_m \cdot \left(\frac{1}{2 \cdot \pi \cdot f_c \cdot (C_z + C_c)} \cdot \frac{\sqrt{1+K^2}}{\sqrt{1+(1/K)^2}} \right) \cdot \frac{1}{15} \cdot \frac{1}{R_{cs}} \cdot A_{ps} = 1$$

Transposing this to find $C_z + C_c$, we get

$$R_s \cdot G_m \cdot \left(\frac{1}{2 \cdot \pi \cdot f_c \cdot} \cdot \frac{\sqrt{1+K^2}}{\sqrt{1+(1/K)^2}} \right) \cdot \frac{1}{15} \cdot \frac{1}{R_{cs}} \cdot A_{ps} = (C_z + C_c)$$

To summarize the previously obtained results, for substitution:

R_{cs} (R1) = 0.18
G_m = 550 μA/V
F_c = 2 kHz
K = 2.077
R_s (R2) = 1.22

The result is:

$$C_z + C_c = 41 \ \text{nF}$$

The fault restart (hiccup) time is about 41 ms if $C_z + C_c = 41$ nF, which is a reasonable value. Earlier we found that $\omega_z = 6{,}050$ and $\omega_p = 26{,}100$, so we can substitute these into the following equation.

$$C_c = (C_z + C_c) \cdot \frac{\omega_z}{\omega_p}$$
$$= 9.5 \ \text{nF}$$

$$C_z = 31.5 \ \text{nF}$$

$$R_z = \frac{1}{\omega_z \cdot C_z}$$
$$= 5.247 \text{ k}\Omega$$

Choose

$$C_c = 10 \text{ nF}, 50 \text{ V}, C0G \text{ capacitor}$$
$$C_z = 33 \text{ nF}, 50 \text{ V}, C0G \text{ capacitor}$$
$$R_z = 5.1k, 1/8 \text{ W}, 1\% \text{ resistor}$$

If the MPS boost controller MP4013B were used instead, calculations would have to be repeated with $G_m = 370$ µA/V. Also because the current sense reference voltages are fixed, new values of R_{cs} (R1) and R_s (R2) would also need to be substituted into the equations. However, a first approximation for the scaling will be approximately 370/550, giving $C_z + C_c = 27$ nF .

By using previous equations and choosing standard component values, we get:

$$C_c = 6.8 \text{ nF}, 50 \text{ V}, C0G \text{ capacitor}$$
$$C_z = 22 \text{ nF}, 50 \text{ V}, C0G \text{ capacitor}$$
$$R_z = 7.5k, 1/8 \text{ W}, 1\% \text{ resistor}$$

6.4.20 Output Clamping Circuit

One problem encountered with a continuous mode boost converter, when operating with $V_{out} < 2 \times V_{in}$, is L–C resonance between the inductor and C_{out}. Clamping the output to the input by a diode from V_{in} to V_{out} can prevent this resonance. This diode is shown as D2 in Fig. 6.10. Diode D2 can be a standard recovery time diode like 1N4002; this type of diode is better at handling surge currents that may occur at initial switch-on.

This completes the design of the HV9912-based boost converter operating in CCM.

6.5 Design of a Discontinuous Conduction Mode Boost LED Driver

As a reminder, discontinuous mode is used when the output voltage is more than 6 times the input voltage. It is also sometimes used in low power applications when the step-up ratio is less than 6 because compensation is simpler.

6.5.1 Design Specification

Input voltage range = 9–16 V

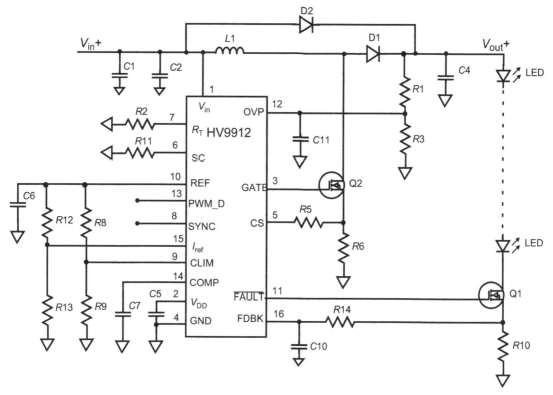

Figure 6.10: Boost Converter With Clamping Diode.

LED string voltage range = 30–70 V

(Note, with a 9-V input and 70-V output, the V_o/V_{in} ratio is approximately 7.8.)

LED current = 100 mA

LED current ripple = 10% (10 mA)

LED dynamic impedance = 55 Ω

Efficiency > 85%

6.5.2 Typical Circuit

A typical circuit for a discontinuous mode boost converter, using the HV9912 boost controller IC is shown in Fig. 6.11. This is similar to the continuous mode circuit shown in Fig. 6.7, but uses a simpler compensation circuit.

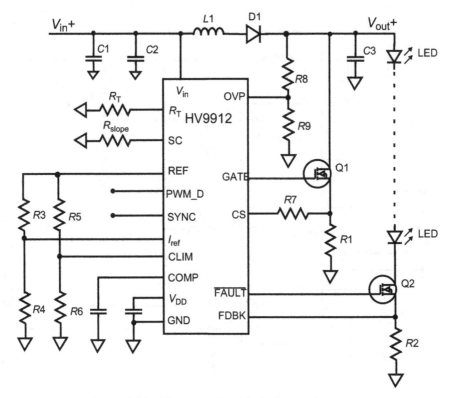

Figure 6.11: Discontinuous Mode Boost Converter.

6.5.3 Selecting the Switching Frequency (f_s)

For low-voltage applications (output voltage < 100 V), and moderate power levels (<30 W), a switching frequency of $f_s = 200$ kHz is a good compromise between switching power loss and size of the components. At higher-voltage or power levels, the switching frequency might have to be reduced to lower the switching losses in the external MOSFET.

6.5.4 Computing the Maximum Inductor Current ($I_{in\,max}$)

The maximum input current is

$$I_{in\,max} = \frac{V_{o\,max} \cdot I_{o\,max}}{\eta_{min} \cdot V_{in\,min}}$$
$$= 0.915\,\text{A}$$

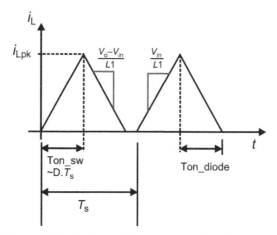

Figure 6.12: Inductor Current Waveform in DCM.

6.5.5 Computing the Input Inductor Value (L1)

Assuming that the sum of the on-time of the switch and the on-time of the diode is 95% of the total switching time period at $V_{\text{in min}}$,

$$L1 \cdot i_{\text{Lpk}} \cdot \left(\frac{1}{V_{\text{in min}}} + \frac{1}{V_{\text{o max}} - V_{\text{in min}}} \right) = \frac{0.95}{f_s}$$
$$= 4.75\,\mu s$$

Where i_{Lpk} is the peak input current (Fig. 6.12).

$V_{\text{in}}/L1$ controls the rate at which current increases and the rising period is determined by the on-time of the MOSFET, which is the duty cycle multiplied by the switching period. The rate of fall is controlled by $(V_o - V_{\text{in}})/L1$ and the falling period is the time that the diode is conducting.

The average input current at the minimum input voltage is equal to the average inductor current and can be computed from

$$I_{\text{in max}} = \frac{1}{2} \cdot i_{\text{Lpk}} \cdot \frac{4.75\,\mu s}{5\,\mu s}$$
$$= 0.475 \cdot i_{\text{Lpk}}$$

Transposing the equation, the peak input current is

$$i_{\text{Lpk}} = \frac{I_{\text{in max}}}{0.475}$$
$$\approx 1.93\,\text{A}$$

Substituting for i_{Lpk} in the equation for $L1$

$$L1 = \frac{0.95}{200k} \cdot \frac{9\,\text{V} \cdot (70\,\text{V} - 9\,\text{V})}{70\,\text{V} \cdot 1.93\,\text{A}}$$
$$= 19.3\,\mu\text{H}$$

Note that the value of $L1$ computed is the absolute maximum value for the inductor. If the inductor value is larger, the inductor current will flow for longer and the circuit will go into CCM. Assuming a ±20% variation in the inductance, the nominal inductor value has to be

$$L1_{\text{nom}} = \frac{L1}{1.2}$$
$$= 16.08\,\mu\text{H}$$

The closest standard value is a 15-μH inductor. This will guarantee DCM operation.

The RMS current through the inductor is

$$I_{\text{Lrms}} = i_{\text{Lpk}} \cdot \sqrt{\frac{0.9}{3}}$$
$$= 1.057\,\text{A}$$

Choose a 15-μH inductor (±20% tolerance). A custom inductor would work best for this application given the large swings in the inductor flux. However, if a standard value inductor is preferred, the saturation current rating of the inductor should be at least 1.5 times the peak current computed, to keep the core losses to an acceptable value. Note that when the circuit is operating at full load, the input current should ramp up linearly. If the rate of current rise increases at the end of the switching cycle, so the ramp is not linear, this indicates that the inductor is beginning to saturate.

The inductor chosen in this case is a 15-μH inductor with an RMS current rating of 1.4 A and a saturation current rating of 3 A.

6.5.6 Computing the On- and Off-Times of the Converter

The on-time of the switch can be computed as

$$t_{\text{on\_sw}} = \frac{L1_{\text{nom}} \cdot i_{\text{Lpk}}}{V_{\text{in\,min}}}$$
$$= 3.22\,\mu\text{s}$$

The on-time of diode is

$$t_{\text{on\_diode}} = \frac{L1_{\text{nom}} \cdot i_{\text{Lpk}}}{V_{\text{o max}} - V_{\text{in min}}}$$
$$= 467 \text{ ns}$$

The maximum duty cycle can then be computed as

$$D_{\text{max}} = t_{\text{on\_sw}} \cdot f_s$$
$$= 0.644$$

The diode conduction time ratio can be expressed as

$$D1 = t_{\text{on\_diode}} \cdot f_s$$
$$= 0.0934$$

6.5.7 Choosing the Switching MOSFET (Q1)

The maximum voltage across the MOSFET in a boost converter is equal to the output voltage. Using a 20% overhead to account to switching spikes, the minimum voltage rating of the MOSFET has to be

$$V_{\text{FET}} = 1.2 \cdot V_{\text{o max}}$$
$$= 84 \text{ V}$$

The RMS current through the MOSFET is

$$I_{\text{FET}} \approx i_{\text{Lpk}} \cdot \sqrt{\frac{D_{\text{max}}}{3}}$$
$$= 0.895 \text{ A}$$

To get the best performance from the converter, the MOSFET chosen has to have a current rating about 3 times the MOSFET RMS current with minimum gate charge Q_g. It is recommended that for designs with the HV9912 because of limited gate drive capability, the gate charge of the chosen MOSFET be less than 25 nC.

The MOSFET chosen for this application is a 100 V, 4.5 A MOSFET with a Q_g of 11 nC.

6.5.8 Choosing the Switching Diode (D1)

The voltage rating of the diode is the same as the voltage rating of the MOSFET (100 V). The average current through the diode is equal to the maximum output current (350 mA). Although the average current through the diode is only 350 mA, the peak current through the diode is equal to i_{Lpk}. Thus, it is a better design approach to choose the current rating of the diode somewhere in between the peak input current and the average output current (preferably closer to the peak input current). Thus, for this design, the diode chosen is a 100 V, 2 A Schottky diode.

As the circuit is operating in discontinuous mode, a fast P–N junction diode could be used. An ultrafast Schottky diode is not required to keep the switching losses low. However, a Schottky diode does have a low-forward voltage drop, so conduction losses are lower that for a P–N junction diode. If cost is more important that efficiency, use a fast P–N junction diode.

6.5.9 Choosing the Output Capacitor (C$_o$)

The value of the output capacitor depends on the dynamic resistance of the LED string, as well as the ripple current desired in the LED string. In designs using the HV9912, a larger-output capacitor (lower-output current ripple) will yield better PWM dimming results. The capacitor value needed to filter the output ripple current will be calculated by considering the RMS current through the LED load.

The output stage of the boost converter is modeled in Fig. 6.13, where the LEDs are modeled as a constant voltage load with series dynamic impedance.

The waveform of the capacitor current in steady state is shown in Fig. 6.14.

Using the 10% peak-to-peak current ripple given in the design parameters table, the maximum voltage ripple across the LED string has to be

$$\Delta v_{p-p} = \Delta I_o \cdot R_{LED}$$
$$= 0.55\,\mathrm{V}$$

Assuming a constant discharging current of 350 mA when the diode current is zero, the equation for the voltage across the capacitor can be written as

$$I_{o\max} = C_o \cdot \frac{\Delta v_{p-p}}{D_{\max} \cdot T_s}$$

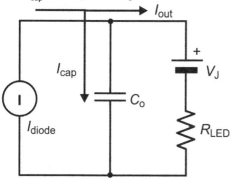

MOSFET off
I_{cap} = short pulse of high current

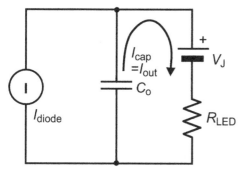

MOSFET on
I_{cap} = long pulse of low current

Figure 6.13: Model of Boost Converter Output.

Figure 6.14: Output Capacitor Current.

Substituting values into this equation,

$$C_o = \frac{I_{o\,max} \cdot D_{max}}{\Delta v_{p-p} \cdot f_s}$$
$$= 0.585\,\mu F$$

The RMS current through the capacitor can be given by

$$I_{rms} = \sqrt{(1 - D1) \cdot I_{o\,max}^2 + \frac{D1}{3} \cdot \left(i_{Lpk} - I_{o\,max}\right)^2}$$
$$= 0.34\,A$$

In this case, a parallel combination of two 1 μF, 100 V metal polypropylene capacitors is chosen.

Note: The proper type of capacitors to use is either metal film capacitors or ceramic capacitors, as they are capable of carrying this high-ripple current. Although ceramic capacitors are smaller in size and capable of carrying the ripple current, they cause a lot of audible noise during PWM dimming. High-value ceramic capacitors are usually limited to 50-V rating. Thus metal polypropylene (or any other metal film) capacitors are the ideal choice for LED drivers if PWM dimming is required.

6.5.10 Choose the Disconnect MOSFET (Q2)

The disconnect MOSFET should have the same voltage rating as the switching MOSFET Q1. The on-state resistance of the MOSFET at room temperature ($R_{on,25°C}$) has to be calculated based on a 1% power loss in Q2 at full-load current. Thus,

$$R_{on,25°C} = \frac{0.01 \cdot V_{omax}}{I_{omax} \cdot 1.4}$$

$$= 5\,\Omega$$

The 1.4 multiplication factor is included to account for the increase in the on-resistance due to rise in junction temperature. In this case, a high Q_g MOSFET can be chosen if desired (as it is not switching regularly), but a high Q_g MOSFET will slow down the turn-on and turn-off times (which might be allowable based on PWM dimming frequency). In this case, the MOSFET chosen is a 100 V, 0.7 Ω, SOT-23 MOSFET with a Q_g of 2.9 nC.

6.5.11 Choosing the Input Capacitors (C1 and C2)

The values of input capacitors C1 and C2 have to be calculated to meet closed-loop stability requirements. The connection from the power source to the boost converter circuit will have some resistance, R_{source}, and some inductance, L_{source}. These feed across the input capacitor (C1 and C2) and so form an LC resonant circuit. To prevent interference with the control loop, the resonant frequency should be arranged to be less than 40% of the switching frequency.

A pair of 22AWG connecting wires 1-ft. (30-cm) long will have an inductance of about 1 μH. This is a good starting point. If necessary, the wires can be twisted together to reduce the inductance.

With a 200-kHz switching frequency, the resonant frequency should be less than 80 kHz.

$$C_{in} \geq \frac{1}{(2 \cdot \pi \cdot f_{LC})^2 \cdot L_{source}} = 3.95\,\mu F$$

C1 = C2 = 2.2 μF, 50 V ceramic.

The maximum source impedance is found using:

$$M = \frac{V_{o\max}}{V_{in\min}} = \frac{70}{9} = 7.778$$

$$R_{source,\max} = \frac{M-1}{M^2 \cdot (M-2)} \cdot R_{LED} = 1.404\,\Omega$$

6.5.12 Calculating the Timing Resistor (R_T)

The HV9912 oscillator has an 18-pF capacitor charged by a current mirror circuit. An external timing resistor R_T provides a reference current for the current mirror. When R_T is connected to 0 V, current flows and the timing process begins. When charged to a certain voltage, the R_S flip-flop is set, the capacitor is discharged, and the timing process starts again. The timing resistor can be calculated by using the following equation:

$$\frac{1}{f_s} \approx R_T \cdot 18\,\text{pF}$$

In this case, for a constant 200-kHz switching frequency, the timing resistor value is approximately 270 kΩ. This resistor needs to be connected between the R_T pin and GND as shown in the typical circuit.

6.5.13 Calculating the Two Current Sense Resistors (R1 and R2)

The value of the output current sense resistor $R2$ can be calculated by limiting its voltage drop to below 0.4 V. Using this criterion,

$$R2 = \frac{0.4\,\text{V}}{I_{o\max}}$$
$$= 4\,\Omega$$

The power dissipation will be 0.4 V $\times$ I$_{o\,\max}$ = 0.04 W. In this case, the resistor chosen is a 3.9 Ω, 1/8 W, 1% resistor.

The MOSFET current sense resistor $R1$ is calculated by limiting the voltage across the resistor to about 250 mV at maximum input current.

$$R1 = \frac{0.25}{i_{L\text{pk}}}$$
$$= 0.12\,\Omega$$

The power dissipated in this resistor is

$$P_{R1} = I_{FET}^2 \cdot R1$$
$$= 0.096\,\text{W}$$

Thus, the chosen current sense resistor is a $0.12\,\Omega$, 1/4 W, 1% resistor.

6.5.14 Selecting the Current Reference Resistors (R3 and R4)

The voltage at the current reference pin I_{REF} can be set either by using the reference voltage provided at the REF pin (through a voltage divider) or with an external voltage source. In the present design, it is assumed that the voltage at the I_{REF} pin is set using a voltage divider from the REF pin. The current reference resistors $R3$ and $R4$ can be computed using the following two equations:

$$R3 + R4 = \frac{1.25\,\text{V}}{50\,\mu\text{A}} \leq 25\,\text{k}\Omega$$

$$\frac{1.25\,\text{V}}{R3 + R4} \cdot R4 = I_{o\,\text{max}} \cdot R2 = 0.1 \cdot 3.9 = 0.39\,\text{V}$$

For this design, the values of the two resistors can be computed to be

$$R3 = 19.1\,\text{k}\Omega,\ 1/8\,\text{W},\ 1\%$$
$$R4 = 8.66\,\text{k}\Omega,\ 1/8\,\text{W},\ 1\%$$

6.5.15 Setting the Inductor Current Limit (R5 and R6)

The inductor current limit value depends on two factors: the maximum inductor current and the slope compensation signal added to the sensed current. Another resistor divider from the REF pin ($R5$ and $R6$) is connected to the CLIM pin and sets the maximum inductor current. The voltage at the CLIM pin can be computed as

$$V_{CLIM} \geq 1.2 \cdot i_{Lpk} \cdot R1$$

This equation assumes that the current limit level is set at about 120% of the maximum inductor current $I_{in\,max}$.

For this design,

$$V_{CLIM} = 1.2 \cdot 1.93 \cdot 0.12$$
$$\geq 0.278\,\text{V}$$

Using a maximum current sourced out of REF pin of 50 μA, the two resistors can be calculated as

$$R5 = 20k, 1/8\,\text{W}, 1\%$$
$$R6 = 6.04k, 1/8\,\text{W}, 1\%$$

No capacitor should be connected at the CLIM pin because this will affect the circuit at start-up.

6.5.16 Capacitors at V_{DD} and REF Pins

It is recommended that bypass capacitors be connected to both V_{DD} and REF pins. For the V_{DD} pin, the capacitor used should be a 10-V ceramic chip capacitor. For low-power designs, a 1-μF capacitor is adequate. If the design uses high-gate charge switching MOSFETs ($Q_g > 15$ nC), the capacitor at the V_{DD} pin should be increased to 2.2 μF.

For the REF pin, the capacitor used is a 0.1-μF ceramic chip capacitor.

6.5.17 Setting the Overvoltage Trip Point (R8 and R9)

The overvoltage trip point can be set at a voltage 15% higher than the maximum steady state voltage. Using a 15% margin, the maximum output voltage during open LED condition will be:

$$V_{open} = 1.15 \cdot V_{o\,max}$$
$$= 80.5\,\text{V}$$

Then, the resistors that set the overvoltage set point can be computed as

$$R8 = \frac{(V_{open} - 5\,\text{V})^2}{0.1}$$
$$= 57\,\text{k}\Omega$$

This equation will allow us to select a 1/8-W resistor by limiting the power dissipation in the resistor.

$$R9 = \frac{R8}{(V_{open} - 5\,\text{V})} \cdot 5\,\text{V}$$
$$= 3.78\,\text{k}\Omega$$

The closest 1% resistor values are

$$R8 = 56k, 1/8\,\text{W}, 1\%$$
$$R9 = 3.6k, 1/8\,\text{W}, 1\%$$

Note: The actual overvoltage point will vary from the desired point by ±5% due to the variation in the reference (see datasheet). For this design, it varies from about 76.5 to 84.5 V. As the maximum output voltage is designed to be 70 V, we have at least 6.5-V margin.

The HV9912 uses a 5-V reference voltage for the OVP comparators, which is why 5 V appears in the equations for R8 and R9. In boost controllers where this reference voltage is different, such as the LTC3783 that uses 1.23 V, this value should be substituted for 5 V.

6.5.18 Designing the Compensation Network

The compensation needed to stabilize the converter could be either a type-I circuit (a simple integrator) or a type-II circuit (an integrator with an additional pole-zero pair). The type of the compensation circuit required will be dependent on the phase of the power stage at the crossover frequency.

The loop gain of the closed-loop system is given by

$$\text{Loop gain} = R_s \cdot G_m \cdot Z_c(s) \cdot \frac{1}{15} \cdot \frac{1}{R_{cs}} \cdot G_{ps}(s)$$

Where G_m is the transconductance of the op-amp (550 µA/V), $Z_c(s)$ is the impedance of the compensation network, and $G_{ps}(s)$ is the transfer function of the power stage. Please note that although the resistors give a 1:14 ratio, the overall effect when including the diode drop is effectively 1:15.

To compute the transfer function for the DCM boost converter in peak current control mode, we need to define a couple of factors.

$$M = \frac{V_{o\max} \cdot I_{o\max}}{V_{o\max} \cdot I_{o\max} - 0.5 \cdot L1_{nom} \cdot i_{L.pk}^2 \cdot f_s}$$

$$M = \frac{70 \cdot 0.1}{70 \cdot 0.1 - 0.5 \cdot 15 \cdot 10^{-6} \cdot 1.93^2 \cdot 200 \cdot 10^3}$$

$$M = \frac{7}{1.41265} = 4.9552$$

$$G_r = \frac{M-1}{2 \cdot M - 1} = \frac{3.95522}{8.9104} = 0.4439$$

For frequencies less than 1/10th of the switching frequency, the power stage transfer function is given by

$$G_{ps}(s) = 2 \cdot \frac{I_{omax}}{i_{Lpk}} \cdot \frac{G_r}{1 + s \cdot R_{LED} \cdot C_o \cdot G_r}$$

$$G_{ps}(s) = 2 \cdot \frac{0.1}{1.93} \cdot \frac{0.4439}{1 + s \cdot 55 \cdot 2 \cdot 10^{-6} \cdot 0.4439} = \frac{0.4439}{1 + s \cdot 48.829 \cdot 10^{-6}}$$

For the present design, choose a crossover frequency $\sim 0.01\, f_s$, or $f_c = 2$ kHz. The low-crossover frequency will result in large values for C_c and C_z, which will indirectly provide a soft-start for the circuit. As the HV9912 does not depend on the speed of the controller circuit for the PWM dimming response, the low-crossover frequency will not have an adverse effect on the PWM dimming rise and fall times. By substituting $s = i \cdot (2\pi \cdot f_c) = i \cdot 12{,}566$ into the transfer function, we get:

$$G_{ps}(s) = \frac{0.046}{1 + s \cdot 0.6136}$$

The magnitude and frequency of the power stage transfer function are:

$$\left\| G_{ps}(s) \right\|_{fc=2\,\text{kHz}} = A_{ps} = 0.039$$

$$\angle G_{ps}(s) \Big|_{fc=2\,\text{kHz}} = \phi_{ps} = -31.5°$$

To get a phase margin of about $\phi_m = 45$ degrees (the recommended phase margin range is 45–60 degrees), the phase boost required will be

$$\phi_{boost} = \phi_m - \phi_{ps} - 90°$$
$$= 45° + 31.5° - 90°$$
$$= -13.5°$$

Based on the value of the phase boost required, the type of compensation can be determined.

$$\phi_{boost} \leq 0° \Rightarrow \text{Type-I controller}$$
$$0° \leq \phi_{boost} \leq 90° \Rightarrow \text{Type-II controller}$$
$$90° \leq \phi_{boost} \leq 180° \Rightarrow \text{Type-III controller}$$

A type-I controller is required to compensate a HV9912 in discontinuous mode. This is much easier to design compared to the type-II compensation described earlier for CCM operation (Table 6.1).

So, for the present design, a simple type-I controller will suffice. The following equation is for the loop gain:

$$R2 \cdot G_m \cdot \left(\frac{1}{2 \cdot \pi \cdot f_c \cdot C_c} \right) \cdot \frac{1}{15} \cdot \frac{1}{R1} \cdot A_{ps} = 1$$

All that is needed is to adjust the gain of the loop to be 1 at the crossover frequency. Transposing this equation, we obtain the value of C_c that will be needed to achieve this.

$$C_c = R2 \cdot G_m \cdot \left(\frac{1}{2 \cdot \pi \cdot f_c} \right) \cdot \frac{1}{15} \cdot \frac{1}{R1} \cdot A_{ps}$$

Reminder of previous results:

$R2$ (R_s) = 3.9
$R1$ (R_{cs}) = 0.12
G_m = 550 µA/V
ω_c = 12,566
A_{ps} = 0.039

So, but substitution:

$$C_c = 3.9 \cdot 550 \cdot 10^{-6} \cdot \left(\frac{1}{12566} \right) \cdot \frac{1}{15} \cdot \frac{1}{0.12} \cdot 0.039 = 3.698 \, \text{nF}$$

Choose C_c = 3.9 nF, 50 V, *C0G* capacitor

This completes our DCM boost converter design using the HV9912. If you are using the LTC3783 boost controller from Linear Technology, G_m = 588 µA/V, so upscaling gives a C_c value of 3.95 nF; hence 3.9 nF could be used, as the HV9912 design. If instead you are using the MP4013B boost controller IC from MPS, G_m = 370 µA/V so the value of C_c should be scaled down accordingly (2.49 nF; use the nearest standard value of 2.2 nF or 2.7 nF).

6.6 Common Mistakes

1. The most common mistake is not having adequate OVP at the output. If the LEDs are disconnected while the circuit is operating, the output voltage will rise until components start to break down. The overvoltage limit set at the output of the boost converter should be lower than the breakdown voltage of any component connected across it. However, in some low-cost, sealed backlight modules, OVP is not provided because the circuit is deemed unrepairable.

2. Testing the circuit with a short string of LEDs. In this case, the forward voltage drop of the LEDs may be lower than the supply voltage, so there will be a high-current flow and the LEDs are likely to be destroyed.
3. Using multilayer ceramic capacitors across the output when PWM dimming will be used. Multilayer ceramic capacitors are piezoelectric, so PWM dimming will cause an audible whistle to be produced.

6.7 Conclusions

Boost converters are used when the minimum output voltage is at least 1.5 times the input voltage. Below this step-up ratio, buck–boost converters should be used instead. These will be described in the next chapter.

CCM is recommended for applications with low input to output step-up ratios, when the output voltage is a maximum of 6 times the input voltage. This keeps the EMI levels relatively low, although care should still be taken in PCB layout and filtering.

DCM is necessary if the output voltage is more than 6 times the input voltage. In some applications, discontinuous mode is used at low step-up ratios because compensation calculations are easier. However, but the EMI produced by a discontinuous mode boost converter is much higher than for a CCM boost converter of similar power output.

Boost–Buck Converter

Chapter Outline

A boost–buck converter allows for the load voltage to be higher or lower than the supply voltage. They are usually used in battery-powered applications because the battery voltage varies widely over time. The most common applications are automotive related: they are used in lighting applications for cars and trucks of course, but they are also used for lighting in caravans, camping, industrial and agricultural machinery (forklift trucks, tractors, diggers, forestry machines, etc.).

Two common types of boost–buck converter are Ćuk and single-ended primary inductance converter (SEPIC), which are single-switch converters. These comprise a boost converter followed by a buck converter. These will be described in this chapter. A less common type is the single inductor fly-back converter, which will be described in Chapter 9 with other fly-back circuits.

The power train of typical boost–buck circuit topology (used as an LED driver) is shown in Fig. 7.1.

The boost–buck converter has many advantages:

- The converter can both boost and buck the input voltage. Thus, it is ideal for cases where the output LED string voltage can be either above or below the input voltage during operation.

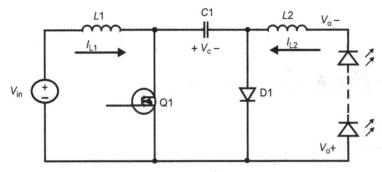

Figure 7.1: Boost–Buck (Ćuk) Power Train.

- The Ćuk converter has inductors in series with both the input and the output. Operating both stages in continuous conduction mode (CCM) will enable continuous currents in both inductors with low current ripple, which greatly reduces the filter capacitor requirements at both input and output. Continuous input current mode reduces the peak current levels and thus helps greatly in meeting conducted electromagnetic interference (EMI) standards.
- All the switching nodes in the Ćuk circuit are isolated between the two inductors. The input and output nodes are relatively quiet. This will minimize the radiated EMI from the converter. With proper layout and design, the converter can easily meet radiated EMI standards.
- One of the advantages of the boost–buck converter is the capacitive isolation. The failure of the switching transistor will short the input and not affect the output. Thus, the LEDs are protected from failure of the MOSFET.
- The two inductors $L1$ and $L2$ can be coupled together on one core. When coupled on a single core, the ripple in the inductor current from one side can be transferred completely to another side (ripple cancellation technique). This would allow, for example, the input ripple to be transferred completely to the output side making it very easy for the converter to meet conducted EMI standards. The circuit must operate in constant frequency mode in order to remain stable with coupled inductors.

Boost–buck circuits are often used in automotive applications, where the LED voltage can be higher or lower than the supply voltage. The battery voltage in a vehicle can vary considerably, depending on whether the battery is charging or discharging, the load on the system, etc. Also, the LED load voltage can change as the LEDs get warm and their forward voltage drops. By boosting first, higher voltage nodes in the circuit are then available to provide power to keep the switching regulator powered when the supply voltage dips to low levels, such as during "cold-crank" (when the motor is being started).

7.1 The Ćuk Converter

Supertex (now part of Microchip) was the first supplier to introduce a dedicated Ćuk converter control IC for LED driving. This is the HV9930, which is a hysteretic mode controller. Hysteretic mode uses the output current level to control the switching, by comparing the voltage across a current sense resistor with an internal reference voltage. Thus the switching frequency changes as the supply and load voltages change. Detailed design procedures will be described in Section 7.1.6. Even readers not using the HV9930 will find useful information within these pages.

Texas Instruments has produced the TPS92690 for Ćuk converter driving. This is intended for coupled inductors, where the input and output inductors use the same magnetic core. In order to allow this, fixed frequency switching is used. The circuit topology allows the LED anode to be connected to ground, since current sensing takes place during the off-time, when the flywheel diode is conducting. A low value resistor in series with the flywheel diode is used for this purpose.

In spite of the many advantages of the Ćuk converter, four disadvantages exist which prevent its widespread use:

- The output is inverted with respect to the input. So the cathode is negative with respect to 0 V/ground. This can cause problems in a few applications.
- The converter is difficult to stabilize. Complex compensation circuitry is often needed to make the converter operate properly. This compensation also tends to slow down the response of the converter, which inhibits the PWM dimming capability of the converter (essential for LEDs).
- An output current–controlled boost–buck converter tends to have an uncontrolled and undamped resonance due to an L–C pair ($L1$ and $C1$). The resonance of $L1$ and $C1$ leads to excessive voltages across the capacitor, which can damage the circuit.
- The input and output inductors, combined with the intermediate and output capacitors, form a series LC circuit. If an input voltage step is applied, via a switch or relay contact, damped ringing can occur with the peak current reaching several amps (I have seen 20 A peaks in a 6-W circuit). Nothing much can be done to reduce this, since the Ćuk controller and the switching MOSFET are not active at this time. Adding resistors into the circuit would help, but at the cost of power dissipation and overall efficiency.

The damping of $L1$ and $C1$ can easily be achieved by adding a damping R–C circuit across $C1$. However, the problem of compensating the circuit so that it is stable is more complex.

7.1.1 Operation of an HV9930 Ćuk Boost–Buck Converter

The Microchip HV9930 solves the problem of compensation and achieving a fast PWM dimming response by using hysteretic current mode control. This ensures fast response and accurate current levels. However, a simple hysteretic current mode control would not work, as the converter would not be able to start-up. To overcome this problem, the HV9930 has two hysteretic current mode controllers—one for the input current and another for the output current.

During start-up, the input hysteretic controller dominates and the converter is in input current limit mode. Once the output current has built up to the required value, the output hysteretic controller takes over. This approach will also help in limiting the input current during start-up (thus providing soft-start); also current is limited in the case of an output overload or input undervoltage condition. Three resistors (for each of the two hysteretic controllers) are required to set both the current ripple and the average current, which enables a simple controller design. Thus six resistors determine the input and output performance.

This section will detail the operation of the boost–buck converter and the design of an HV9930-based converter. The design example is specifically designed for automotive applications, but it can also be applied for any DC/DC applications. At the time of writing, there is only one other device with the same functionality as the HV9930, which is the AT9933. The AT9933 has an automotive temperature specification (up to 125°C operation), whereas the HV9930 is intended for industrial and commercial applications.

The diagram of the power train for a Ćuk boost–buck converter was shown previously, in Fig. 7.1.

In steady state, the average voltages across both $L1$ and $L2$ are zero. Thus, the voltage, V_c, across the middle capacitor $C1$ is equal to the sum of the input and output voltages.

$$V_c = V_{in} + V_o$$

When switch Q1 is turned on, the currents in both inductors start ramping up (Fig. 7.2).

$$L1\frac{di_{L1}}{dt} = V_{in}$$

$$L2\frac{di_{L2}}{dt} = V_c - V_o = V_{in}$$

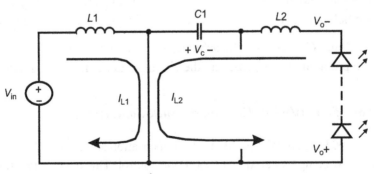

Figure 7.2: Ćuk Circuit: MOSFET On.

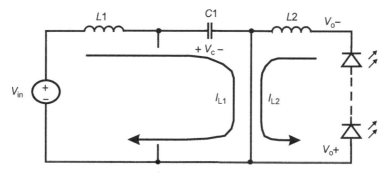

Figure 7.3: Ćuk Circuit: MOSFET Off.

When switch Q1 is turned off, the currents in both inductors start ramping down (Fig. 7.3).

$$L1\frac{di_{L1}}{dt} = V_{in} - V_c = -V_o$$

$$L2\frac{di_{L2}}{dt} = -V_o$$

Assuming that the switch is ON for a duty cycle D and using the fact that, in steady state, the total volt-seconds applied across any inductor is zero, we get

$$V_{in} \cdot (D) = V_o \cdot (1 - D)$$

$$\Rightarrow \frac{V_o}{V_{in}} = \frac{D}{1 - D}$$

Thus, the voltage transfer function obtained for the boost–buck converter will give buck operation for $D < 0.5$ and boost operation for $D > 0.5$. The steady state waveforms for the converter are shown in Fig. 7.4.

The maximum voltage seen by Q1 and D1 is equal to the voltage across the capacitor $C1$.

$$V_{Q1} = V_{D1} = V_c$$

The standard boost–buck converter is modified, by adding three additional components, for proper operation of the HV9930 (Fig. 7.5).

A damping circuit R_d–C_d has been added to damp the $L1$–$C1$ pair. These additional components stabilize the circuit. Typically C_d is ten times the value of $C1$ and R_d is a few ohms.

An input diode (D2) has been added. This diode is necessary for PWM dimming operation (in case of automobile applications, this could be the reverse polarity protection diode). This diode helps to prevent capacitors $C1$ and C_d from discharging when the gate signals for Q1

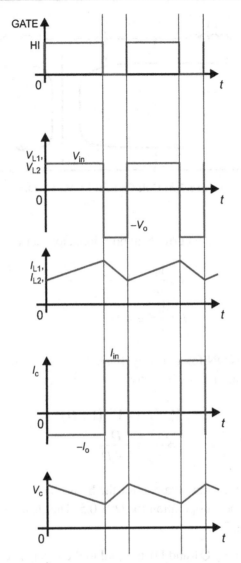

Figure 7.4: Ćuk Converter Steady State Waveforms.

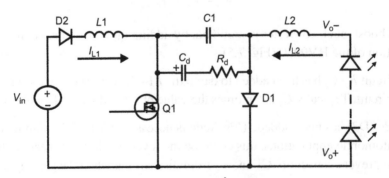

Figure 7.5: Modified Ćuk Circuit.

are turned off. Thus, when the HV9930 is enabled, the steady-state output current level will be reached quickly.

7.1.2 Hysteretic Control of the Ćuk Converter

Hysteretic control keeps the current in inductor L2 (i_{L2}) between preset upper and lower boundaries. As previously shown in Fig. 7.4, the inductor current ramps up at a rate of $V_{in}/L2$ when the switch is ON and ramps down at a rate of $-V_o/L2$ when the switch is OFF. Thus, the hysteretic control scheme turns the switch OFF when the inductor current reaches the upper limit and turns the switch ON when it reaches the lower limit.

The average current in inductor L2 is then set at the average of the upper and lower thresholds. The ON and OFF times (and thus the switching frequency) vary as the input and output voltages change to maintain the inductor current levels. However, in any practical implementation of hysteretic control, there will be comparator delays involved. The switch will not turn ON and OFF at the instant the inductor current hits the limits, but after a small delay time, as illustrated in Fig. 7.6.

7.1.3 The Effects of Delay in Hysteretic Control

This delay time introduces two unwanted effects:

- It alters the average output current value. For example, if the delay on the downslope of the inductor current is more than the delay on the upslope, then the average current value decreases.
- It decreases the switching frequency, because the turn-on period and the turn-off period are both extended by the delays, which may make it more difficult for the circuit to meet EMI regulations.

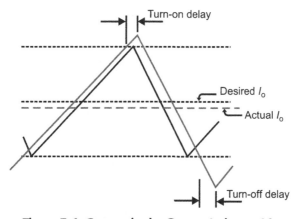

Figure 7.6: Current in the Output Inductor L2.

These effects will have to be taken into consideration when choosing the output inductor value and setting the current limits.

Assume a peak-to-peak current ripple setting of Δi_o (using the programming resistors) and a desired average current I_o. A hysteretic current-controlled boost–buck converter acts as a constant-off-time converter as long as the output voltage is fixed, and the off-time is theoretically independent of the input voltage. Thus, the converter is designed assuming a constant off time T_{off} (the method to determine the off-time will be discussed later in the application note).

For the HV9930, as long as the switching frequencies are less than 150 kHz, these delay times have a negligible effect and can be ignored. In these cases, the output inductor can be determined by

$$L2 = \frac{V_o \cdot T_{off}}{\Delta i_o}$$

If the inductor chosen is significantly different from the computed value, the actual off-time $T_{off,ac}$ can be recomputed using the same equation.

However, in automotive applications, there is an advantage in setting the switching frequency of the converter below 150 kHz or in the range between 300 and 530 kHz. This will place the fundamental frequency of the conducted and radiated EMI outside of the restricted bands making it easier for the converter to pass automotive EMI regulations. In cases where the switching frequency is more than 300 kHz, the delay times cannot be neglected and have to be accounted for in the calculations. Fig. 7.7 illustrates the output inductor current waveform and the various rise and fall times.

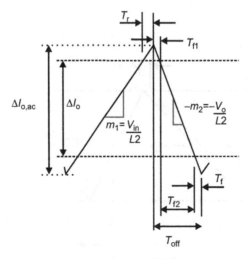

Figure 7.7: Hysteretic Control With Comparator Delays.

From Fig. 7.7,

$$T_{\text{off}} = T_{f1} + T_{f2} + T_f$$

$$= \frac{V_{\text{in}}}{V_o} \cdot T_r + T_{f2} + T_f$$

The desired output current ripple Δi_o and the downslope of the inductor current m_2 determine T_{f2}. The delay times of the HV9930 determine T_r and T_f. For the HV9930, the delay time of the comparators is related to the overdrive voltage (voltage difference between the two input terminals of the current sense comparator) applied as

$$T_{\text{delay}} \approx \frac{K}{\sqrt[3]{m \cdot (0.1/\Delta i_o)}}$$

where "m" is the rising or falling slope of the inductor current.

$$T_r = \frac{6\mu}{\sqrt[3]{\dfrac{V_{\text{in}} \cdot 0.1}{\Delta i_o}}} \cdot \sqrt[3]{L2} = K_1 \cdot \sqrt[3]{L2}$$

$$T_{f2} = \frac{\Delta i_o \cdot L2}{V_o} = K_2 \cdot L2$$

$$T_f = \frac{6\mu}{\sqrt[3]{\dfrac{V_o \cdot 0.1}{\Delta i_o}}} \cdot \sqrt[3]{L2} = K_3 \cdot \sqrt[3]{L2}$$

To find the value of $L2$ using the previously mentioned time delay equations results in a cubic equation. This cubic has one real root and two complex roots. The inductor value is the real root of the cubic raised to the third power.

$$a = K_2$$

$$b = \frac{V_{\text{in}}}{V_o} \cdot K_1 + K_3$$

$$c = T_{\text{off}}$$

$$\Delta = 12 \cdot \sqrt{3} \cdot \sqrt{\frac{4 \cdot b^3 + 27 \cdot a \cdot c^2}{a}}$$

$$L2 = \left\{ \frac{1}{6 \cdot a} \left[(108 \cdot c + \Delta) \cdot a^2 \right]^{1/3} - \frac{2 \cdot b}{\left[(108 \cdot c + \Delta) \cdot a^2 \right]^{1/3}} \right\}^3$$

The actual off-time $T_{\text{off,ac}}$ can be computed by substituting the chosen inductor value back into the equations for T_r, T_f, and T_{f2}, to get $T_{r,\text{ac}}$, $T_{f,\text{ac}}$, and $T_{f2,\text{ac}}$.

$$T_{\text{off,ac}} = T_{f1,\text{ac}} + T_{f2,\text{ac}} + T_{f,\text{ac}}$$
$$= \frac{V_{\text{in}}}{V_o} \cdot T_{r,\text{ac}} + T_{f2,\text{ac}} + T_{f,\text{ac}}$$

The actual ripple in the inductor current $\Delta i_{o,\text{ac}}$ is

$$\Delta i_{o,\text{ac}} = \frac{V_o \cdot T_{\text{off,ac}}}{L2}$$

7.1.4 Stability of the Ćuk Converter

The single-switch Ćuk converter can be considered as separate boost and buck converters (in that order), which are cascaded, and both switches being driven with the same signal (Fig. 7.8).

The relationships between the voltages in the system are

$$\frac{V_c}{V_{\text{in}}} = \frac{1}{1-D} \quad \text{(boost converter)}$$
$$\frac{V_o}{V_c} = D \quad \text{(buck converter)}$$

The capacitor voltage V_c and the input–output relationship can both be derived using the previous equations

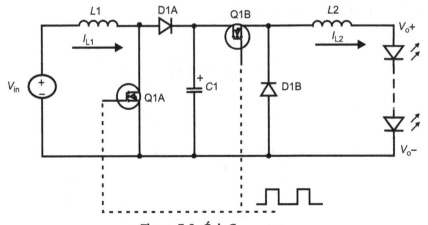

Figure 7.8: Ćuk Converter.

$$\frac{V_{o}}{V_{in}} = \frac{V_{o}}{V_{c}} \cdot \frac{V_{c}}{V_{in}} = \frac{D}{1-D}$$

$$V_{c} = \frac{V_{in}}{1-D} = \frac{V_{in}}{1-(V_{o}/V_{c})}$$

$$\Rightarrow V_{c} = V_{o} + V_{in}$$

For the purposes of designing the damping network, it is easier to visualize the converter in its two-switch format of Fig. 7.7 rather than as the single-switch Ćuk converter. Hence, for the remainder of this section, the cascaded converter will be used to derive the equations.

In hysteretic control of the Ćuk converter using the HV9930, once the start-up phase is complete, the output buck stage is controlled and the input boost stage is uncontrolled. An equivalent schematic of the HV9930-controlled Ćuk converter is shown in Fig. 7.9.

The hysteretic control of the buck stage ensures that the output current i_{L2} is constant under all input transient conditions. So, for the purposes of average modeling, the load seen by the capacitor C1 can be modeled as a current source equal to $d \cdot I_{o}$, where d is the instantaneous duty cycle and I_{o} is the constant output current. The CCM buck stage also imposes one more constraint:

$$V_{o} = d \cdot v_{c}$$

where d and v_{c} are the time-dependent duty cycle and capacitor voltage and V_{o} is the constant output voltage. For the system to be stable, it is necessary that the control system will act to reduce any disturbance in capacitor voltage.

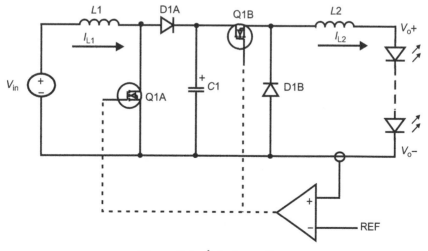

Figure 7.9: Ćuk Controller.

The loop gain of the system for a boost–buck converter without damping has a negative phase margin (i.e., the phase is less than $-180°$ when the magnitude crosses 0 dB). This is due to the undamped LC pole pair and causes the system to be unstable. Thus, any disturbance to the capacitor voltage will get amplified and keep increasing till the components breakdown.

The addition of R–C damping of this undamped pole pair can stabilize the system and make sure that the disturbance input is properly damped. Also, the presence of C_d ensures that R_d will not see the DC component of the voltage V_c across it, reducing the power dissipated in the damping resistor (C_d blocks the DC component of the voltage).

Assuming $C_d \gg C1$, the loop gain transfer function of the R–C-damped boost–buck converter can be derived as

$$G(s)H(s) = \frac{D}{1-D} \cdot \frac{\left(1+s \cdot R_d \cdot C_d\right) \cdot \left(1 - s \cdot \dfrac{D}{(1-D)^2} \cdot \dfrac{L1 \cdot I_o}{V_o}\right)}{\left(1+s \cdot R_d \cdot C1\right) \cdot \left(1+s \cdot R_d \cdot C_d + s^2 \cdot \dfrac{L1 \cdot C_d}{(1-D)^2}\right)}$$

Thus, the loop has a DC gain of $D/(1-D)$ and includes:

1. Damping (and ESR) zero at $\omega_z = \dfrac{1}{R_d \cdot C_d}$;

2. RHP zero at $\omega_{RHP} = \dfrac{(1-D)^2}{D} \cdot \dfrac{V_o}{L1 \cdot I_o}$;

3. Complex double pole with natural resonant frequency $\omega_o = \dfrac{1-D}{\sqrt{L1 \cdot C_d}}$ and damping factor $\delta = (1-D) \cdot R_d \cdot \sqrt{\dfrac{C_d}{L1}}$;

4. High-frequency pole at $\omega_p = \dfrac{1}{R_d \cdot C1}$.

In order to achieve stable loop, the 0 dB crossing (ω_c) must be placed such that $\omega_c \ll \omega_{RHP}$ and $\omega_c \ll \omega_p$. The latter condition is easily met by selecting $C_d \gg C1$.

We can easily obtain approximate values of C_d and R_d for the case of $\omega_c \gg \omega_o$. This condition is usually met for the worst-case calculations at minimum input voltage, since the DC gain is the highest at this condition. Set $\omega_c = \omega_{RHP}/N$, where $N \gg 1$. Then ω_o can be approximately calculated from

$$\omega_o = \omega_c \cdot \sqrt{\frac{1-D}{D}} = \frac{\omega_{RHP}}{N} \cdot \sqrt{\frac{1-D}{D}}$$

Substituting for ω_0 and ω_{RHP} into the above gives the equation for computing C_d:

$$C_d = \frac{N^2 \cdot D^3}{(1-D)^3} \frac{L1 \cdot I_o^2}{V_o^2}.$$

Selecting R_d such that $\omega_z = \omega_c$ results in a good phase margin with minimum power dissipation. Then, using equations for ω_z and ω_{RHP} gives a solution for R_d:

$$R_d = \frac{N \cdot D}{(1-D)^2} \frac{L1 \cdot I_o}{C_d \cdot V_o}.$$

Using the previous equations, the approximate values for the damping network can be computed using the following equations:

$$C_d = 9 \cdot \left(\frac{D}{1-D}\right)^3 \cdot L1 \cdot \left(\frac{I_o}{V_o}\right)^2$$

$$R_d = \frac{3 \cdot D}{(1-D)^2} \cdot \frac{L1 \cdot I_o}{C_d \cdot V_o}$$

Note that the damping resistor value includes the ESR or the damping capacitor. In many cases, the damping capacitor is chosen to be an electrolytic capacitor, which will have a significant ESR. In such cases, the damping resistor can be reduced accordingly.

7.1.5 Dimming Ratio Using PWM Dimming

The linearity in the dimming ratio achievable with the Boost–Buck depends on both the switching frequency and the PWM dimming frequency.

For a converter designed to operate at a minimum switching frequency of 300 kHz, one switching time period equals 3.33 μs. This is the minimum on-time of the PWM dimming cycle. At a PWM dimming frequency of 200 Hz (5-ms period), 3.33 μs equals a minimum duty cycle of 0.067%. This corresponds to a 1:1500 dimming range. However, the same converter being PWM dimmed at 1 kHz (1-ms time period) will have a minimum duty ratio of 0.33% or a PWM dimming range of 1:300.

If the minimum on-time of the PWM dimming cycle is less than the switching time period, the LED current will not reach its final value. Hence the average current will be less. Thus, the LEDs will dim, but there will be a loss of linearity between the average LED current and the duty cycle of the PWM input.

7.1.6 Design of the Ćuk Converter With HV9930

7.1.6.1 Specification

Input voltage: 9–16 V (13.5 V typical)

Transient voltage: 42 V (clamped load dump rating)

Reverse polarity protection: 14 V

Output voltage: 28 V maximum

Output current: 350 mA

LED resistance: 5.6 Ω

Estimated efficiencies: 72% minimum, 82% maximum (80% typical)

These efficiency values do not take into account the power loss in the reverse blocking diode. The diode will drop about $V_d = 0.5\,V$ across it and thus will dissipate power in the range of 0.4–0.6 W. This diode voltage drop will be taken in account while designing the converter.

The efficiency values used in this design are typical values for the given input voltages and output power level. Higher efficiencies can be obtained at lower input current levels (i.e., higher input voltages): the efficiency drop at lower input voltages is due to conduction losses caused by the correspondingly larger input currents. The efficiency values will depend on the operating conditions and, except in very high power designs, these values can be used as a good approximation.

Efficiencies higher than 85% can easily be achieved with the HV9930-controlled Ćuk converter if the operating frequency is kept below 150 kHz. However because of automotive EMI requirements, the higher efficiencies are traded-off for higher switching frequencies (which increase switching losses in the system).

Consider a Ćuk converter circuit as shown in Fig. 7.10.

7.1.6.2 Switching frequency at minimum input voltage

Although the HV9930 is a variable frequency IC, the selection of the minimum switching frequency is important. In the case of automotive converters, designing with a switching frequency in the range between 300 and 530 kHz would avoid the main radio broadcast bands and make it easier to meet the conducted and radiated EMI specifications. So, for this application we choose a minimum switching frequency of 300 kHz (which occurs at minimum input voltage).

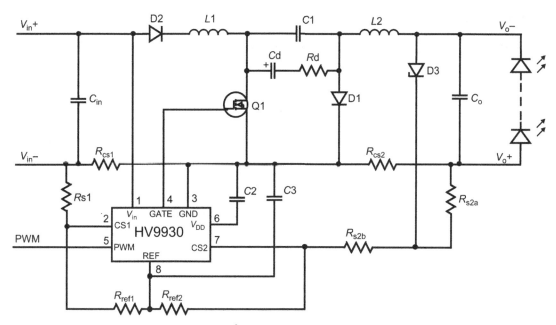

Figure 7.10: Ćuk Converter Using HV9930.

7.1.6.3 Calculating the duty cycle

The switch duty cycle will have to be computed at the minimum input voltage.

$$D_{max} = \frac{1}{1 + \dfrac{\eta_{min} \cdot \left(V_{in,min} - V_d\right)}{V_o}}$$

$$= 0.821$$

7.1.6.4 Calculating the input current

The input current level at the minimum input voltage should be calculated first because this gives the highest current level. The value obtained will be used to work out the current ratings of the various components.

$$I_{in,max} = \frac{V_o \cdot I_o}{\eta_{min} \cdot \left(V_{in,min} - V_d\right)}$$

$$= 1.601 \text{ A}$$

7.1.6.5 Calculating the output inductor

The first step is to compute the off-time. The off-time of the converter can be calculated as

$$T_{off} = \frac{1-D_{max}}{f_{s,min}}$$
$$= 598 \text{ ns}$$

Assuming a 25% peak-to-peak ripple in the output current ($\Delta i_o = 87.5$ mA), and accounting for the diode drop in the input voltage by substituting $V_{in,min}-V_d$ in place of V_{in}, yields

$$598 \text{ ns} = 0.887 \ \mu \cdot \sqrt[3]{L2} + 3.125 \text{ m} \cdot L2 + 1.89 \ \mu \cdot \sqrt[3]{L2}$$

Solving for $L2$ gives

$$L2 = (0.052)^3 = 145 \ \mu H$$

The closest standard value inductor is a 150 µH, 0.35 A RMS, and 0.4 A sat inductor.

Since the inductance value is different from the computed value, the actual off-time will also change as

$$T_{off,ac} = 2.777 \ \mu \cdot \sqrt[3]{L_{2,ac}} + 3.125 \text{ m} \cdot L_{2,ac}$$
$$= 616 \text{ ns}$$

The actual ripple in the output current is given by

$$\Delta i_{o,ac} = \frac{V_o \cdot T_{off,ac}}{L_{2,ac}}$$
$$= 0.115 \text{ A}$$

Note that although the ripple in the output current was assumed to be about 25% (or 87.5 mA), the actual ripple is almost double that value. This increase in the ripple is due to the delays of the comparators. A capacitor will be required at the output of the converter (across the LEDs) to reduce the ripple to the desired level. This capacitor will be very small, as the switching frequencies are large, but the capacitor will also help to reduce output EMI.

It is also useful to calculate the ripple overshoot and undershoot beyond the programmed limits. This will help determine how the average current changes due to the delays.

$$\Delta i_{over} = \frac{V_o}{L_{2,ac}} \cdot \left(\frac{V_{in,min} - V_d}{V_o} \cdot K_1 \right) \cdot \sqrt[3]{L_{2,ac}}$$
$$= 8.3 \text{ mA}$$

$$\Delta i_{under} = \frac{V_o}{L_{2,ac}} \cdot K_3 \cdot \sqrt[3]{L_{2,ac}}$$
$$= 19 \text{ mA}$$

Thus, the average output current will be reduced from the set value by about 10.7 mA.

In most cases, due to the inductor values available, the actual off-time will differ from the computed value significantly. Thus, it is better to use the actual value of the off-time calculated in order to work out the rest of the component values.

If the switching frequency is less than 150 kHz, the equation $L2 = \dfrac{V_o \cdot T_{off}}{\Delta i_o}$ can be used to calculate the output inductance (*L2*) value, simplifying the procedure greatly.

7.1.6.6 Calculating the input inductor

We can assume a 15% peak-to-peak ripple in the input current at minimum input voltage (this low input ripple will minimize the input filtering capacitance needed). The off-time previously calculated can be used to find the value of the input inductor.

$$L1 = \frac{V_o \cdot T_{off,ac}}{0.15 \cdot I_{in,max}}$$
$$= 72 \ \mu H$$

The closest standard value inductor is an 82 μH inductor. The current rating of this inductor will be decided in the final stages after the input current limit has been set.

The peak-to-peak ripple in the input current is

$$\Delta I_{in} = \frac{V_o \cdot T_{off,ac}}{L_{1,ac}}$$
$$= 0.21 \ A$$

7.1.6.7 Calculating the value of the middle capacitor (C1)

Assuming a 5% ripple across the capacitor at minimum input voltage ($\Delta v_c = 0.1 \cdot (V_{in,min} - V_d + V_o) = 1.825 \ V$), capacitor *C1* can be calculated as

$$C1 = \frac{I_{in,max} \cdot T_{off,ac}}{\Delta v_c}$$
$$= 0.54 \ \mu F$$

$$I_{rms,C1} = \sqrt{I_{in,max}^2 \cdot (1 - D_{max}) + I_o^2 \cdot D_{max}}$$
$$= 0.72 \ A$$

The voltage rating and type of this capacitor have to be chosen carefully. This capacitor carries both the input current and the output current. Thus, to prevent excessive losses and overheating of the capacitor, it must have a very low ESR. Ceramic capacitors are an ideal choice for this application due to their low ESR and high transient voltage limit. If a ceramic capacitor cannot be used for reasons of cost or availability, a plastic film capacitor, such as PET can be used instead.

The maximum steady state voltage across the capacitor is 44 V (= 28 V + 16 V), and the maximum transient voltage across the capacitor $V_{c,max}$ is 70 V (= 28 V + 42V). Ceramic capacitors can easily withstand up to 2.5 times their voltage rating for the duration of the load dump voltage. Also, the actual capacitance value of these capacitors reduces based on the bias voltage applied. Ceramic capacitor types X7R and X5R are more stable and the capacitance drop is not more than 20% at full rated voltage.

Thus, a 0.47 μF, 50 V X7R ceramic chip capacitor can be selected.

7.1.6.8 Choosing the switching transistor (Q1)

The peak voltage across the MOSFET Q1 is 70 V. Assuming a 30% overhead on the voltage rating to account for leakage inductance spikes, the MOSFET voltage needs to be at least

$$V_{FET} = 1.3 \cdot V_{c,max}$$
$$= 91 \text{ V}$$

The RMS current through the MOSFET will be at maximum level at low input voltage (higher current levels and maximum duty cycle). The maximum RMS current through the MOSFET is

$$I_{FET,max} = (I_{in,max} + I_o) \cdot \sqrt{D_{max}}$$
$$= 1.77 \text{ A}$$

A typical choice for the MOSFET is to pick one whose current rating is about 3 times the maximum RMS current. Choose FDS3692 from Fairchild Semiconductors (100 V, 4.5 A, 50 mΩ N-channel MOSFET). Note that the current rating is normally quoted at 25°C.

The total gate charge Q_g of the chosen MOSFET is a maximum of 15 nC. It is recommended that the MOSFET total gate charge should not exceed 20 nC, as the large switching times will cause increased switching losses. A higher gate charge would be allowable if the switching frequency can be reduced appropriately.

7.1.6.9 Choosing the switching diode

The maximum voltage rating of the diode D2 is the same as the MOSFET voltage rating. The average current through the diode is equal to the output current.

$$I_{diode} = I_o = 350 \text{ mA}$$

Although the average current of the diode is only 350 mA, the actual switching current through the diode goes as high as 1.95 A ($I_{in,max} + I_o$). (Note: the calculations were for

360 mA, to allow for 10 mA drop because of delays, but the actual average current is 350 mA.) A 500-mA diode will be able to carry the 1.79 A current safely, but the voltage drop at such high current levels would be extremely large increasing the power dissipation. Thus, we need to choose a diode whose current rating is at least 1 A. A 100 V, 2 A Schottky diode would be a good choice.

7.1.6.10 Choosing the input diode

The input diode serves two purposes:

1. It protects the circuit from a reverse polarity connection at the input.
2. It helps in PWM dimming of the circuit by preventing $C1$ from discharging when the HV9930 is turned off.

The current rating of the device should be at least equal to $I_{in,max}$. The voltage rating of the device should be more than the reverse input voltage rating. A higher current rating often gives a lower forward voltage drop. In this case, a 30BQ015 (15 V, 3 A Schottky diode) would be a good choice.

7.1.6.11 Calculating the input capacitance

Some capacitance is required on the input side to filter the input current. This capacitance is mainly responsible for reducing the second harmonic of the input current ripple (which in this case falls in the AM radio band). According to the SAE J1113 specifications, the peak limit for narrowband emissions in this range is 50 dBμV to meet Class 3 at an input voltage of 13 ± 0.5 V. Assuming a sawtooth waveform for the input current as a conservative approximation, the RMS value of the second harmonic component of the input current ($I_{in,2}$) can be computed as

$$I_{in,2} = \frac{\Delta I_{in}}{2 \cdot \sqrt{2} \cdot \pi} = 0.024 \text{ A}$$

The switching frequency of the converter at 13 V input can be computed as

$$D_{nom} = \frac{1}{1 + \dfrac{\eta_{nom} \cdot (V_{in,nom} - V_d)}{V_o}}$$

$$= \frac{1}{1 + \dfrac{0.8 \cdot (13.5 - 0.5)}{28}}$$

$$= 0.73$$

$$f_{s,nom} = \frac{1 - D_{nom}}{T_{off,ac}}$$

$$= 414 \text{ kHz}$$

$$C_{in} = \frac{I_{in,2}}{4 \cdot \pi \cdot f_{s,nom} \cdot 10^{-6} \cdot 10^{50/20}}$$

$$= 14.6 \, \mu F$$

Choose a parallel combination of three 4.7 µF, 25 V, X7R ceramic capacitors.

7.1.6.12 Calculating the output capacitance

The output capacitance is required to reduce the LED current ripple from 115 mA to ΔI_{LED} = 70mA (20% peak-to-peak ripple) can be approximately calculated by using only the first harmonic in the inductor current. A 70 mA peak-to-peak ripple in the LED results in a 392 mV ($\Delta v_o = \Delta I_{LED} \cdot R_{LED}$) peak-to-peak ripple voltage. Then

$$\frac{\Delta v_o}{2} = \frac{8}{\pi^2} \cdot \left(\frac{\Delta i_{L2}}{2} \right) \cdot \frac{R_{LED}}{\sqrt{1 + \left(2 \cdot \pi \cdot f_{s,min} \cdot R_{LED} \cdot C_o \right)^2}}$$

The output capacitance required can then be calculated from this as

$$C_o = \frac{\sqrt{\left(\frac{16 \cdot R_{LED}}{\pi^2} \cdot \frac{\Delta i_{L2}}{\Delta v_o} \right)^2 - 1}}{2 \cdot \pi \cdot f_{s,min} \cdot R_{LED}}$$

$$= 0.178 \, \mu F$$

Use a 0.22 µF, 35 V ceramic capacitor.

7.1.6.13 Calculating the theoretical switching frequency variation

The maximum and minimum frequencies (using steady state voltage conditions) can be now be worked out:

$$f_{s,min} = \frac{1 - \dfrac{1}{1 + \eta_{min} \cdot (V_{in,min} - V_d)/V_o}}{T_{off,ac}}$$

$$= 291 \, kHz$$

$$f_{s,max} = \frac{1 - \dfrac{1}{1 + \eta_{max} \cdot (V_{in,max} - V_d)/V_o}}{T_{off,ac}}$$

$$= 506 \, kHz$$

The theoretical frequency variation for this design is 398 kHz ± 27%.

7.1.6.14 Design of the damping circuit

The values for the damping network can be calculated as follows

$$C_d = 9 \cdot \left(\frac{D_{max}}{1 - D_{max}} \right)^3 \cdot L_{1,ac} \cdot \left(\frac{I_o}{V_o} \right)^2$$
$$= 11 \, \mu F$$

$$R_d = \frac{3 \cdot D_{max}}{(1 - D_{max})^2} \cdot \frac{L_{1,ac} \cdot I_o}{C_d \cdot V_o}$$
$$= 7.16 \, \Omega$$

The power dissipated in R_d can be computed as

$$P_{R_d} = \frac{\Delta v_c^2}{12 \cdot R_d}$$
$$= \frac{3.65^2}{12 \cdot 7.16} = 0.155 \, W$$

The RMS current through the damping capacitor will be

$$i_{C_d} = \frac{\Delta v_c}{2 \cdot \sqrt{3} \cdot R_d} = 0.147 \, A$$

Choose a 10 μF, 50 V electrolytic capacitor that can allow at least 150 mA RMS current. An example would be EEVFK1H100P from Panasonic (10 μF, 50 V, Size D). This capacitor has about a 1 Ω ESR, so R_d can be reduced to about 6.2 Ω.

7.1.6.15 Internal voltage regulator of the HV9930

The HV9930 includes a built-in 8–200 V linear regulator. This regulator supplies the power to the IC. This regulator can be connected at either one of two nodes on the circuit as shown in Fig. 7.11.

In the normal case, when the input voltage is always greater than 8 V, the V_{in} pin of the IC can be connected to the cathode of the input protection diode (as shown in Fig. 7.11A). If reverse protection is not provided, the V_{in} pin can be connected directly to the positive supply.

In conditions where the converter needs to operate at voltages lower than 8 V, once the converter is running (as in the case of cold-crank operation), the VIN pin of the HV9930 can be connected as shown in Fig. 7.11B. In this case, the drain of the MOSFET is at $V_{in} + V_o$, and hence even if the input voltage drops below 8 V, the IC will still be functioning. However,

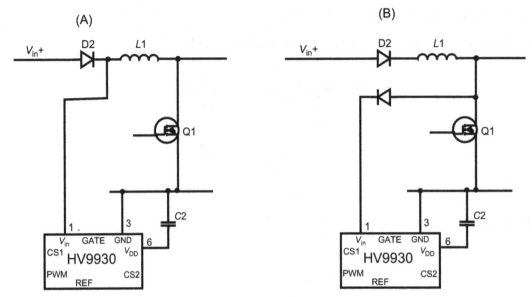

Figure 7.11: Connection Points for V_{in}. (A) HV9930 powered directly from supply. (B) Powered via switching node.

in this case, more hold-up capacitance will be required at the V_{DD} pin to supply the power to the IC when the MOSFET is ON.

In both cases, a 2.2 µF or greater value ceramic capacitor is recommended at the V_{DD} pin.

7.1.6.16 Internal voltage reference

The HV9930 includes an internal 1.25 V (±3%) reference. This reference can be used to set the current thresholds for the input and output hysteretic comparators. It is recommended that this pin be bypassed with at least a 0.1 µF ceramic capacitor, in order to keep the impedance low and reduce noise.

7.1.6.17 Programming the hysteretic controllers and overvoltage protection

The input and output current levels for the hysteretic controllers are set by means of three resistors for each current—one current sense resistor and two divider resistors. The equations governing the resistors are the same for both the input and output sides and are given as

$$\frac{R_s}{R_{ref}} = \frac{0.05 \cdot \dfrac{\Delta i}{I} + 0.1}{1.2 \cdot \dfrac{\Delta i}{I} - 0.1}$$

$$R_{cs} = \frac{1.2 \cdot \dfrac{R_s}{R_{ref}} - 0.05}{I}$$

These equations assume that the 1.25 V reference provided by the HV9930 is used to set the current. In cases where linear dimming of the LEDs is required, it is recommended that the input current thresholds be based on the 1.25 V reference and the output current thresholds are modified using the variable input voltage available. In such a case, assuming the maximum external voltage V_{LD} as the reference, the previous two equations can be modified as

$$\frac{R_s}{R_{ref}} = \frac{0.05 \cdot \frac{\Delta i}{I} + 0.1}{(V_{LD} - 0.05) \cdot \frac{\Delta i}{I} - 0.1}$$

$$R_{cs} = \frac{(V_{LD} - 0.05) \cdot \frac{R_s}{R_{ref}} - 0.05}{I}$$

In this design example, it is assumed that linear dimming is not required and the 1.25 V reference is used for both the input and output programming.

Note: The HV9930 cannot operate the boost–buck converter in the discontinuous conduction mode, so there is a minimum external voltage reference that can be used. The minimum external voltage is given by

$$V_{LD} = 0.1 \cdot \frac{R_{ref2} + R_{s2}}{R_{s2}},$$

The programming of the output side is also linked to the overvoltage protection. The boost–buck converter is not inherently programmed against open LED conditions, so external protection is required. This is achieved by adding Zener diode D3, and by splitting the resistor R_{s2} into two parts—R_{s2a} and R_{s2b}. In normal operation, the inductor current will flow only through R_{cs2} and the voltage drop across R_{cs2} is sensed through R_{s2a} and R_{s2b} in series.

When there is an open LED condition, the inductor current will flow through diode D3. This will then clamp the output to the Zener breakdown voltage. However, since the diode cannot take the full design current, the current level has to be reduced to more manageable levels. During open LED conditions, the current will flow though both R_{cs2} and R_{s2a}. Thus, the effective current sense resistor seen by the IC is $R_{cs2} + R_{s2a}$ and the voltage drop across both of these will be sensed through R_{s2b}. This provides a feedback signal to the output comparator to reduce the programmed current level and thus prevent the high LED currents from flowing into the Zener diode.

7.1.6.18 Choosing the output side resistors

For the output current, $I_o = 0.36$A (to compensate for the 10 mA drop due to the delay times) and $\Delta I_o = 87.5$ mA. Note that we are using the values assumed and not the actual values computed in Section 7.1.6.5 for the ripple current. Using these values in the previous equations,

$$\frac{R_{s2a} + R_{s2b}}{R_{ref2}} = 0.534$$

$$R_{cs2} = 1.64 \ \Omega$$
$$P_{R_{cs2}} = 0.35^2 \cdot 1.64 = 0.2 \ \text{W}$$

Before we complete the design of the output side, we also have to design the overvoltage protection. For this application, choose a 33-V Zener diode. This is the voltage at which the output will clamp in case of an open LED condition. For a 350-mW diode, the maximum current rating at 33 V works out to be about 10 mA. Using a 5 mA current level during open LED conditions, and assuming the same R_s / R_{ref} ratio,

$$R_{s2a} + R_{cs2} = 120 \ \Omega.$$

Choose the following values for the resistors:

$$R_{cs2} = 1.65 \ \Omega, 1/4 \ \text{W}, 1\%$$
$$R_{ref2} = 10 \ \text{k}\Omega, 1/8 \ \text{W}, 1\%$$
$$R_{s2a} = 100 \ \Omega, 1/8 \ \text{W}, 1\%$$
$$R_{s2b} = 5.23 \ \text{k}\Omega, 1/8 \ \text{W}, 1\%$$

7.1.6.19 Design of the input side resistors

For the input side, we first have to determine the input current level for limiting. This current level is dictated by the fact that the input comparator must not interfere with the operation of the circuit, even at minimum input voltage.

The peak of the input current at minimum input voltage will be

$$I_{in,pk} = I_{in,max} + \frac{\Delta I_{in}}{2}$$
$$= 1.706 \ \text{A}$$

Assuming a 30% peak-to-peak ripple when the converter is in input current limit mode, the minimum value of the input current will be

$$I_{lim,min} = 0.85 \cdot I_{in,lim}$$

We need to ensure that $I_{\text{lim,min}} > I_{\text{in,pk}}$ for proper operation of the circuit. Assuming a 5% safety factor, that is,

$$I_{\text{lim,min}} = 1.05 \cdot I_{\text{in,pk}},$$

we can compute the input current limit to be $I_{\text{in,min}} = 2.1\,\text{A}$.

Allowing for a 30% peak-to-peak ripple, we can calculate

$$\frac{R_{\text{s1}}}{R_{\text{ref1}}} = 0.442$$

$$R_{\text{cs1}} = 0.228\ \Omega$$
$$P_{R_{\text{cs1}}} = I_{\text{in,lim}}^{2} \cdot R_{\text{cs1}} = 1\ \text{W}$$

This power dissipation is a maximum value, which occurs only at minimum input voltage. At a nominal input voltage of 13.5 V, we can compute the input current using the nominal values for the efficiency and the input voltage.

$$I_{\text{in,nom}} = \frac{28 \cdot 0.35}{0.8 \cdot (13.5 - 0.5)}$$
$$= 0.942\ \text{A}$$

$$P_{R_{\text{cs1}}} = 0.942^{2} \cdot 0.228 = 0.2\ \text{W}$$

Thus, at nominal input voltage, the power dissipation reduces by about 5 times to a reasonable 0.2 W.

Choose the following values for the resistors:

$$R_{\text{cs1}} = \text{Parallel combination of three } 0.68\ \Omega, 1/2\ \text{W}, 5\%\ \text{resistors}$$

$$R_{\text{ref1}} = 10\ \text{k}\Omega, 1/8\ \text{W}, 1\%$$
$$R_{\text{s1}} = 4.42\ \text{k}\Omega, 1/8\ \text{W}, 1\%$$

7.1.6.20 *Input inductor current rating*

The maximum current through the input inductor is $I_{\text{lim,max}} = 1.15 \cdot I_{\text{in,lim}} = 2.4$ A. Thus, the saturation current rating of the inductor has to be at least 2.5 A. If the converter is going to be in input current limit for extended periods of time, the RMS current rating needs to be 2 A, else a 1.5 A RMS current rating will be adequate.

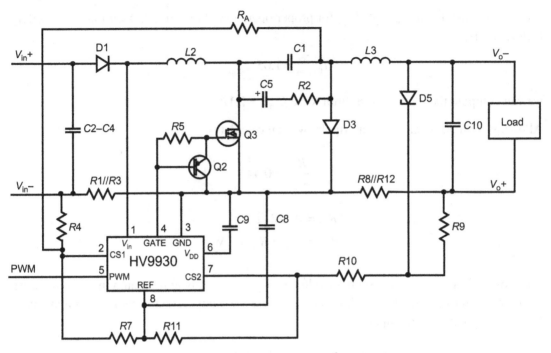

Figure 7.12: Modification of the Ćuk Circuit.

7.1.6.21 Improving efficiency

The input current sense resistor can be reduced in value, which gives reduced power
dissipation (loss). To allow this, it is necessary to add an extra resistor (R_A) between the anode
of the flywheel diode and the current sense input of the HV9930 (AT9933). This resistor
allows a reduction in the hysteresis required by the input comparator. The additional resistor
is shown in Fig. 7.12.

In Fig. 7.12, $R_{s1} = R4$, $R_{ref1} = R7$, and $R_{cs1} =$ the parallel combination $R1 \parallel R3$.

Consider the circuit during the period when the MOSFET is ON, so that the input current
through $L1$ is increasing by $\Delta I_{in}/2$, until it reaches $I_{in,lim} + \Delta I_{in}/2$. With the MOSFET turned
ON, the positive side of the capacitor $C1$ is grounded and the other side of $C1$, which is
connected to resistor R_A, is at potential $-V_{C1}$. Note that the potential $-V_{C1,nom} = V_{in,nom} + V_o$.
The voltage reference for the comparator input at CS1 is 0 V. Now consider the node at CS1
in terms of current flow; CS1 is high impedance input, so the sum of currents equal zero:

$$\frac{V_{ref}}{R7} = \frac{V_{C1,nom}}{R_A} + \frac{\left(I_{in,lim} + \dfrac{\Delta I_{in}}{2}\right) \cdot R1 \parallel R3}{R4}$$

Now consider the circuit when the MOSFET is OFF. Now the flywheel diode D3 is conducting, so the negative side of capacitor C1 is grounded (the small forward voltage of the diode can be ignored). With the MOSFET turned OFF, the voltage reference for the comparator input at CS1 is 100 mV.

$$\frac{V_{ref} - 0.1\ \text{V}}{R7} = \frac{0.1\ \text{V}}{R_A} + \frac{0.1\ \text{V} + \left(I_{in,lim} - \dfrac{\Delta I_{in}}{2} \right) \cdot R1 \| R3}{R4}$$

Since R_A is a very large value and the voltage across it is small, we can ignore its effect to simplify the calculations:

$$\frac{V_{ref} - 0.1\ \text{V}}{R7} = \frac{0.1\ \text{V} + \left(I_{in,lim} - \dfrac{\Delta I_{in}}{2} \right) \cdot R1 \| R3}{R4}$$

We can thus ignore the addition of R_A during the period that the MOSFET is turned OFF. Clearly, the value of $R1 \| R3$ can be reduced if the current $\left(I_{in,lim} - \dfrac{\Delta I_{in}}{2} \right)$ can be increased, or if $R4$ can be decreased, or both.

The maximum current sense voltage occurs when the MOSFET is first turned ON.

$$V_{sense,max} = \left(I_{in,lim} + \frac{\Delta I_{in}}{2} \right) \cdot R1 \| R3$$

This is a function of the voltage across the capacitor C1. If we take another look at the equation for current flow when the MOSFET is turned ON:

$$\frac{V_{ref}}{R7} = \frac{V_{C1,nom}}{R_A} + \frac{\left(I_{in,lim} + \dfrac{\Delta I_{in}}{2} \right) \cdot R1 \| R3}{R4}$$

In a Ćuk topology, $V_{C1} = V_{in} + V_{out}$. At start-up, $V_{out} = 0$ V, so $V_{C1,min} = V_{in,min}$. The highest input current occurs at $V_{in,min}$.

$$\frac{V_{ref}}{R7} = \frac{V_{in,min}}{R_A} + \frac{\left(I_{in,lim} + \dfrac{\Delta I_{in}}{2} \right) \cdot R1 \| R3}{R4}$$

If we set the maximum current $\left(I_{in,lim} + \dfrac{\Delta I_{in}}{2} \right)$ in the modified circuit to be equal to the inductor $L1$ saturation current, I_{sat}, we get

$$\frac{V_{ref}}{R7} = \frac{V_{in,min}}{R_A} + \frac{I_{sat} \cdot R1 \| R3}{R4}$$

In practice we start with the design of an unmodified circuit, so $\left(I_{in,lim} + \dfrac{\Delta I_{in}}{2} \right)$ are the values calculated before the addition of R_A is considered. In the modified circuit, I_{sat} (of $L1$) must be much higher than these values to gain the loss reduction benefit, which gives a higher input ripple at start-up.

$$R_A = \frac{\left(V_{in,nom} + V_{out} \right) - \dfrac{V_{in,min} \cdot \left(I_{in,lim} + \dfrac{\Delta I_{in}}{2} \right)}{I_{sat}}}{\dfrac{V_{refl}}{R7} \cdot \left(1 - \dfrac{\left(I_{in,lim} + \Delta I_{in} \right)}{I_{sat}} \right)}$$

$$R4(mod) = \frac{0.1\,\text{V}}{\dfrac{V_{refl} - 0.1\,\text{V}}{R7} - \dfrac{\left(I_{in,lim} - \dfrac{\Delta I_{in}}{2} \right)}{I_{sat}} \cdot \left(\dfrac{V_{refl}}{R7} - \dfrac{V_{in,min}}{R_A} \right)}$$

$$R1 \| R3(mod) = \frac{R4(mod)}{I_{sat}} \cdot \left(\frac{V_{refl}}{R7} - \frac{V_{in,min}}{R_A} \right)$$

$V_{refl} = 1.25$ V in the standard configuration.

Let us see the effect of allowing 30% input current ripple.

$$\Delta I_{in} = 0.3 \cdot I_{in,lim}$$
$$R_A = 1.48\ \text{M}\Omega \to 1.5\ \text{M}\Omega$$
$$R4(R_{s1}) = 2.306\ \text{k}\Omega \to 2.4\ \text{k}\Omega$$
$$R1/R3(R_{cs1}) = 0.091\ \Omega \to 0.09\,\Omega, \quad R1 = R3 = 0.18\ \Omega$$

The input current sense resistor, R_{cs1}, is reduced from 0.228 to 0.09 Ω, which reduces the conduction losses in the circuit to less than half the original value.

7.1.6.22 Meeting conducted and radiated EMI

Due to the nature of the Ćuk boost–buck converter, it is easy to meet conducted and radiated EMI specifications. A few precautions need to be taken during design and PCB layout to be able to meet the EMI standards.

1. In some cases, when the input current ripple is too large or the switching frequency of the converter is above 150 kHz, it might not be possible to meet the conducted EMI standards using only capacitors at the input. In such cases, an input PI filter might be required to filter the low frequency harmonics.

2. Shielded inductors or toroidal inductors should always be preferred over unshielded inductors. These inductors will minimize radiated magnetic fields.
3. During layout, the IC and MOSFET ground connection should be connected to a copper plane on one of the PCB layers with the copper plane extending under the inductors.
4. The loop consisting of Q1, C1, and D1 should be as small as possible. This would help greatly in meeting the high frequency EMI specifications.
5. The length of the trace from GATE output of the HV9930 to the GATE of the MOSFET should be as small as possible, with the source of the MOSFET and the GND of the HV9930 being connected to the GND plane. A low value resistor (10–33 Ω) in series with GATE connection will slow down the switching edges and greatly reduce EMI, although this will cause efficiency to decrease slightly.
6. An R–C damping network might be necessary across diode D1 to reduce ringing due to the undamped junction capacitance of the diode. Such a circuit is often called a "snubber."

This concludes the Ćuk converter design. We can now consider a closely related circuit; the SEPIC.

7.2 SEPIC Boost–Buck Converters

The abbreviation SEPIC comes from the description—single-ended primary inductance converter. A SEPIC is a boost–buck converter, like a Ćuk, so its input voltage range can overlap the output voltage. SEPIC circuits can be designed for constant voltage or constant current output.

The SEPIC topology has been known for some time, but only recently has there been a revival in its application because: (1) it needs low-ESR capacitors and these are now widely available and (2) it can be used to create AC input power supplies with power factor correction (PFC) that are used to meet worldwide EMI standards.

In automotive and portable applications, batteries are used as a power source for DC–DC converters. A 12 V supply used in automotive applications can have a wide range of terminal voltage, typically 9–16 V during normal operation using a lead-acid battery, but can go as low as 6.5 V during cold-crank and as high as 90 V during load-dump (when the battery is disconnected). The peak voltage is usually clamped to about 40 V, using a voltage-dependent resistor to absorb the energy.

Lithium batteries have been very successful in portable applications, thanks mostly to their impressive energy density. A single lithium cell provides an open voltage of 4.2 V when fully charged, and replaces up to three of the alternative NiCd or NiMH cells. During discharge the cell still retains some energy down to 2.7 V. This input voltage range can be both above and below the output voltage of many DC/DC converters and so discounts the possibility of using simple boost or buck converters.

International standards for AC mains power supplies rated above 75 W require PFC. Having a good power factor means that the current waveform from the AC line is sinusoidal and in phase with the voltage. A power factor of 1 is perfect and is only achieved with a resistive load. Most PFC circuits use a simple step-up converter as the input stage, implying that the input stage output must exceed the peak value of the input waveform. In Europe AC inputs of 190–265 V RMS are found, which impose an output of at least 375 V, forcing the following converters to work with elevated input voltages. Typically a PFC input stage has a 400 V output.

By using a SEPIC topology, which has a boost–buck topology, the boost section provides PFC and the buck section produces a lower output voltage. This provides a compact and efficient design. It provides the required output level even if the peak input voltage is higher. More detail on PFC circuits will be given in Chapter 8.

7.2.1 Basic SEPIC Equations

The boost or step-up topology, as shown in Fig. 7.13, is the basis for the SEPIC converter. The boost converter principle is well understood: first, switch Q1 conducts during the on-period, T_{on}, which increases the current in $L1$ and thus increases the magnetic energy stored there. Second, the switch stops conducting during the off-period, T_{off}, but the current through $L1$ cannot change abruptly—it continues to flow, but now through diode D1 and into C_{out}. The current through $L1$ decreases slowly as the stored magnetic energy decreases. Capacitor C_{out} filters the current pulse that was generated by $L1$ when Q1 turned off.

The diode D1 has to switch very quickly, so a diode with a short reverse recovery time (T_{rr} less than 75 ns) is needed. In cases where V_{out} is relatively low, the efficiency can be improved by using a Schottky diode with low forward voltage (about 400 mV) for D1.

Note that a boost converter has one major limitation: V_{out} must always be higher than V_{in}. If V_{in} is ever allowed to become greater than V_{out}, D1 will be forward biased and nothing can prevent current flow from V_{in} to V_{out}.

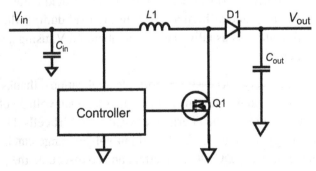

Figure 7.13: The Boost Converter Topology Which is the Basis for SEPIC Power Supply Circuits.

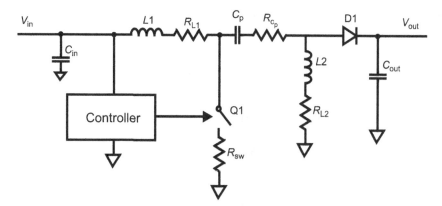

Figure 7.14: SEPIC Topology.

The SEPIC scheme in Fig. 7.14 removes this limitation by inserting a capacitor (C_p) between $L1$ and D1. This capacitor blocks any DC component between the input and output. The anode of D1, however, must connect to a known potential. This is accomplished by connecting D1 to ground through a second inductor ($L2$). $L2$ can be separate from $L1$ or wound on the same core, depending on the needs of the application.

If $L1$ and $L2$ are wound on the same core, which is simply a transformer, one might argue that a classical fly-back topology is more appropriate. However, the transformer leakage inductance, which is not a problem in SEPIC schemes, often requires a snubber network in fly-back schemes. Snubber networks are described later in this chapter; put simply they require additional components that must be carefully selected to minimize losses.

Parasitic resistances that cause most of the conduction losses in a SEPIC are R_{L1}, R_{L2}, R_{sw}, and R_{C_p} which are associated with $L1$, $L2$, SW, and C_p, respectively. These parasitic components are also shown in Fig. 7.14.

An advantage of the SEPIC circuit, besides buck and boost capability, is a capacitor (C_p) that prevents unwanted current flow from V_{in} to V_{out}. Thus the limitation of the simple boost converter, that V_{in} had to always be less than V_{out}, has been overcome.

Though it has very few elements, the operation of a SEPIC converter is not so simple to describe by equations; some assumptions have to be made. First, assume that the values of current and voltage ripple are small with respect to the DC components. Second, assume that at equilibrium there is no DC voltage across the two inductances $L1$ and $L2$ (neglecting the voltage drop across their parasitic resistances). By using these assumptions, C_p sees a DC potential of V_{in} at one side (through $L1$) and ground on the other side (through $L2$). The DC voltage across C_p is:

$$V_{C_p}(\text{mean}) = V_{in}$$

The period of one switching cycle is $T = 1/\text{frequency}$. The portion of T for which switch Q1 is closed is the duty cycle, D, and the remaining part of the period is thus $1 - D$. As the mean voltage across L1 equals zero during steady-state conditions, the voltage seen by L1 during $D * T$ (i.e., the MOSFET "ON" period) is exactly compensated by the voltage seen by L1 during $(1 - D) * T$ (i.e., the MOSFET "OFF" period):

$$D \cdot T \cdot V_{in} = (1-D) \cdot T \cdot (V_{out} + V_D + V_{C_p} - V_{in})$$

where V_D is the forward voltage drop of D1 for a direct current of $(I_{L1} + I_{L2})$, and V_{C_p} is equal to V_{in}. Simplifying this we get:

$$D \cdot T \cdot V_{in} = (1-D) \cdot T \cdot (V_{out} + V_D)$$

Transposing this, we get:

$$\frac{(V_{out} + V_D)}{V_{in}} = \frac{D}{1 - D} = A_i$$

A_i is called the amplification factor, where "i" represents the ideal case for which parasitic resistances are null. Neglecting V_D with respect to V_{out} (as a first approximation), we see that the ratio of V_{out} to V_{in} can be greater than or less than 1, depending on the value of D (with equality obtained for $D = 0.5$).

The more accurate expression A_a (amplification, actual) accounts for parasitic resistances in the circuit:

$$A_a = \frac{V_{out} + V_D + I_{out} \cdot (A_i \cdot R_{C_p} + R_{L2})}{V_{in} - A_i \cdot I_{out} \cdot (R_{L1} + R_{sw}) - R_{sw} \cdot I_{out}}$$

This formula allows computation of the minimum, typical, and maximum amplification factors for V_{in} (A_{a_min}, A_{a_typ}, and A_{a_max}). The formula is recursive ("A_{a_xxx}" appears in both the result and the expression), but a few iterative calculations lead to the solution. The expression neglects switching losses due to the switch Q1 and reverse recovery current in D1. Those losses are usually negligible, especially if Q1 is a fast MOSFET and its drain-voltage swing ($V_{in} + V_{out} + V_d$) remains under 30 V.

In some cases, you should also account for losses due to the reverse recovery current of D1, and for core losses due to high-level swings in stored magnetic energy. You can extrapolate the corresponding values of D:

$$D = A_a/(1 + A_a)$$

Or more generally:

$D\_xxx = A_{a\_xxx}/(1 + A_{a\_xxx})$, where xxx is min, typ, or max.

The DC current through C_p is zero, so the mean output current can only be supplied by L2:

$$I_{out} = I_{L2}$$

The power dissipation requirement for L2 is eased because the mean current into L2 always equals I_{out} and does not depend on variations of V_{in}.

To calculate the current into L1 (I_{L1}), we can use the fact that no DC current can flow through C_p. Thus, the coulomb charge flowing during $D * T$ is perfectly balanced by an opposite coulomb charge during $(1 - D) * T$. When the switch is closed (for an interval $D \cdot T$) the potential at the switch node is fixed at 0 V. Since the capacitor C_p was previously charged to voltage V_{in}, the anode of D1 will now have a potential of $-V_{in}$, which reverse biases D1. Current through C_p is then I_{L2}. When the switch is open during $(1 - D) * T$, current I_{L2} flows through D1 while I_{L1} flows through C_p:

$$D \cdot T \cdot I_{L2} = (1 - D) \cdot T \cdot I_{L1}$$

Knowing that $I_{L2} = I_{out}$,

$$I_{L1} = A_{a\_xxx} \cdot I_{out}$$

Input power equals output power divided by efficiency, so I_{L1} depends strongly on V_{in}. For a given output power, I_{L1} increases if V_{in} decreases. Knowing that I_{L2} (and hence I_{out}) flows into C_p during $D * T$, we choose C_p so that its ripple ΔV_{C_p} is a very small fraction of V_{C_p} (gamma = 1–5%). The worst case occurs when V_{in} is minimal.

$$C_p \rangle \frac{I_{out} \cdot D_{min} \cdot T}{\text{gamma} \cdot V_{in\_min}}$$

By using a high switching frequency, small multilayer ceramic capacitors can be used for C_p. However, ensure that C_p is able to sustain the power dissipation $\left(P_{C_p}\right)$ due to its own internal equivalent series resistance $\left(R_{C_p}\right)$:

$$P_{C_p} = A_{a\_min} \cdot R_{C_p} \cdot I_{out}^2$$

The MOSFET switch drain-to-source resistance, in series with a current sense resistor for limiting the maximum current, is given by the term R_{sw}. This incurs the following loss:

$$P_{sw} = A_{a\_min} \cdot (1 + A_{a\_min}) \cdot R_{sw} \cdot I_{out}^2$$

Losses $P_{R_{L1}}$ and $P_{R_{L2}}$, due to the internal resistances of L1 and L2, are easily calculated:

$$P_{R_{L1}} = A_{a\_min}^2 \cdot R_{L1} \cdot I_{out}^2$$
$$P_{R_{L2}} = R_{L2} \cdot I_{out}^2$$

When calculating the loss due to D1, the average power loss is due to the output current and the forward voltage drop of D1:

$$P_{D1} = V_D \cdot I_{out}$$

L1 is chosen so its total current ripple (ΔI_{L1}) is a fraction ($\beta = 20\text{--}50\%$) of I_{L1}. The worst case for β occurs when V_{in} is at maximum because ΔI_{L1} is at maximum when I_{L1} is at minimum. Assuming $\beta = 0.5$:

$$L1\_min = \frac{2 \cdot T \cdot (1 - D_{max}) \cdot V_{in\_max}}{I_{out}}$$

Choose a standard value nearest to that calculated for L1, and make sure its saturation current meets the following condition:

$$I_{L1-sat} \gg I_{L1} + 0.5 \cdot \Delta I_{L1} = \frac{A_{a\_min} \cdot I_{out} + 0.5 \cdot T \cdot D_{min} \cdot V_{in\_min}}{L1}$$

The calculation for L2 is similar to that for L1:

$$L2\_min = \frac{2 \cdot T \cdot D_{max} \cdot V_{in\_max}}{I_{out}}$$

$$I_{L2-sat} \gg I_{L2} + 0.5 \cdot \Delta I_{L2} = \frac{I_{out} + 0.5 \cdot T \cdot D_{max} \cdot V_{in\_max}}{L2}$$

If the windings of L1 and L2 are wound on the same core, you must choose the larger of the two inductance values when calculating the number of turns required. The two windings should be wound bifilar (twisted around each other before being wound on the core) and thus will have the same number of turns and the same inductance values. Otherwise, voltages across the two windings will differ and C_p will act as a short circuit to the difference. If the winding voltages are identical, they generate equal and additive current gradients. In other words, there will be mutual inductance of equal value in both windings. Thus, the inductance measured across each isolated winding (when there is nothing connected to the other winding) should equal only half of the value calculated for L1 and L2.

As no great potential difference exists between the two windings, you can save costs by winding them together in the same operation. If the windings' cross-sections are equivalent, the resistive losses will differ because their currents (I_{L1} and I_{L2}) differ. Total loss, however, is lowest when losses are distributed equally between the two windings, so it is useful to set each winding's cross-section according to the current it carries. This is particularly easy to do when the windings consist of insulated strands of wire (Litz) for counteracting the skin effect. Finally, the core size is chosen to accommodate a saturation current much greater than ($I_{L1} + I_{L2} + \Delta I_{L1}$) at the highest core temperature anticipated.

The purpose of the output capacitor (C_{out}) is to average the current pulses supplied by D1 during T_{off}. The capacitor must be able to handle high-level repetitive surge currents with low ESR and low self-inductance. Fortunately, ceramic and plastic film capacitors meet these requirements. The minimum value for C_{out} is determined by the amount of ripple (ΔV_{out}) that can be tolerated:

$$C_{out} \geq \frac{A_{a\_min} \cdot I_{out} \cdot D_{min} \cdot T}{\Delta V_{out}}$$

The actual value of the output capacitor may need to be much larger than that calculated using the previous equation, especially if the load current is composed of high-energy pulses. The input capacitor can be very small, thanks to the filtering properties of the SEPIC topology. Usually, C_{in} can be one-tenth the value of C_{out}:

$$C_{in} = C_{out}/10$$

Overall efficiency η can be predicted from V_{in} and A_a. The result can be misleading because it doesn't account for the switch-transition losses or core losses and the real efficiency could be much lower:

$$\eta = V_{out}/A_a V_{in}$$

Finally, the switch SW and diode D1 should be rated for breakdown voltages with a 15% margin:

$$V_{DS}(\text{switch}) > 1.15 \ (V_{out} + V_D + V_{in})$$
$$V_R(\text{diode}) > 1.15 \ (V_{out} + V_{in})$$

Example:

Let $V_{in} = 50 - 150$ V and $V_{out} = 15$ V at 1 A maximum. Let us operate at 200 kHz switching frequency, so that $T = 5$ μs. Now $\dfrac{V_{out}}{V_{in}} = \dfrac{D}{1-D}$, so $D_{max} = 0.231$ and $D_{min} = 0.091$.

$L1_{min} = 2T(1 - D_{max})V_{in\_max}/I_{out}$

$L1_{min} = 10^{-5} * 0.769 * 150/1 = 1.15$ mH; let $L1 = 1.5$ mH

$L2_{min} = 2TD_{max}V_{in\_max}/I_{out}$

$L2_{min} = 10^{-5} * 0.231 * 150/1 = 0.347$ mH; let $L2 = 0.47$ mH

$C_p > I_{out} \cdot D_{min}T/(\text{gamma} \cdot V_{in\_min})$

$C_p > 1 * 0.091 * 2 * 10^{-5}/(0.05 * 50) = 728$ nF; let $C_p = 1$ μF.

Now $D_{xxx} = A_{a\_xxx}/(1 + A_{a\_xxx})$, where xxx is min, typ, or max. So A_{a_min} occurs at $D_{min} = 0.091$ and $A_{a\_min} = 0.1$.

$$C_{out} \geq A_{a\_min} \cdot I_{out} \cdot D_{min} \cdot T/\Delta V_{out}$$

$C_{out} > 0.1 * 1 * 0.091 * 2 * 10^{-5}/0.1$.

$C_{out} >> 1.82\ \mu F$. Let $C_{out} = 100\ \mu F$.

$C_{in} > C_{out}/10$. Let $C_{in} = 10\ \mu F$.

So, the fundamental component values have been calculated. Now what remains for the designer is the choice of suitable (and available) parts.

7.2.2 SEPIC Control Integrated Circuits

Suitable control circuits are parts designed for boost converters. These include: the Microchip's HV9911/HV9912/HV9963, Linear Technology's LTC3783, Texas Instruments' LM3421, and ON Semiconductor's NCP3065. Those with the highest input voltage rating are the HV9911 (200 V) and the HV9912 (90 V). These are more suitable for trains, trucks, and industrial vehicles (like forklift tractors and digging machines), where higher battery voltages are used.

A schematic of a SEPIC circuit using a coupled inductor is given in Fig. 7.15. This circuit uses a Texas Instruments' controller, the LM3421, and full details are given in the application note AN-2009 "LM3421 SEPIC LED Driver Evaluation Board for Automotive Applications."

You will notice that both D1 and Q1 have a series RC circuit connected in parallel. This is a snubber to help reduce EMI. During fast switching, oscillations will be caused by parasitic capacitance and inductance in series. The snubber passes AC signals because of the series capacitance, but the signal amplitude will be damped by the series resistance. Snubbers can cause losses and add cost, so they are tend to be used only when necessary. The circuit described in Fig. 7.15 uses snubbers because it is for automotive lighting; the EMI limits for automotive are low level and difficult to achieve.

7.3 Buck–Boost Topology

Unlike the boost–buck circuits used by the Ćuk and SEPIC topologies, the buck–boost uses a single inductor. It is a fly-back circuit and hence will be covered in Chapter 9.

7.4 Four-Switch Buck–Boost

Some constant voltage buck–boost ICs use four MOSFET switches and a single inductor. For example, Intersil's ISL9120. I have not seen this topology used in LED drivers so far, but it can be efficient and only uses one inductor (Fig. 7.16).

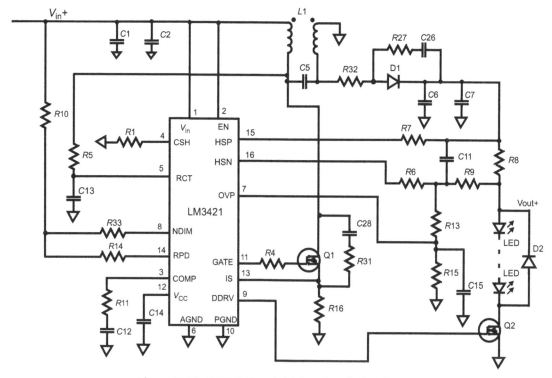

Figure 7.15: SEPIC Circuit Using Coupled Inductor.

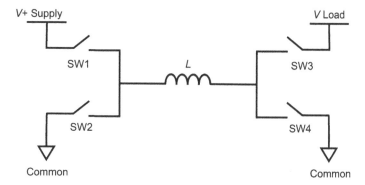

Figure 7.16: Four-Switch Buck–Boost Topology.

In buck mode, switches SW1 and SW2 are controlled by the regulator IC, SW3 is ON and SW4 is OFF. In boost mode, SW1 is ON, SW2 is OFF and switches SW3 and SW4 are controlled by the regulator IC. When the input voltage is approximately equal to the desired output voltage, switch SW1 and SW3 are both ON and switches SW2 and SW4 are both OFF. Instead of feedback from a regulated voltage output, it should be possible to regulate output current by using the voltage developed across a current sense resistor in series with LEDs.

7.5 Common Mistakes in Boost–Buck Circuits

Most mistakes concern the choice of inductors.

1. Using too low inductance value. Boost–buck circuits operate with both inductors in CCM. Hence the inductor should be chosen with an inductance value higher than that calculated, to allow for tolerances and for saturation effects (the inductance falls with increasing current). Calculate the value, add 20% tolerance, and then pick the next highest standard value.
2. Using an inductor with too low saturation current rating. Note that inductor manufacturers give two current ratings. The saturation current rating (I_sat) is the current level that causes a certain drop in measured inductance; this is usually for a 10% drop, but check this because some manufacturers quote for 20% inductance drop. This current rating is very important in switching converters because of the high peak current levels that can be present. The RMS current rating (I_r, or I_rms) is given for a certain temperature rise in the core, typically 40°C. Its value depends on the winding resistance and the thermal characteristics of the materials used in the inductor assembly. This current rating is measured without any regard to the inductance, so can be higher or lower than I_{sat}.

7.6 Conclusions

The boost–buck is an ideal topology where the LED load voltage can be higher or lower than the supply voltage. It should also be used when the supply voltage is not more than 20% different (worst case) from the LED load voltage. So if the LED voltage (maximum) is 20 V and the supply voltage (minimum) is 23 V, the difference is 3 V, and 3/20 = 0.15 or 15%, so a Ćuk or SEPIC should be used. If the supply voltage is always more than 20% higher than the load voltage, use a buck topology. If the supply voltage is always more than 20% lower than the load voltage, use a boost topology. The boost–buck is less efficient compared to buck or boost topologies and is usually more costly.

CHAPTER 8

Nonisolated Power Factor Correction Circuits

Chapter Outline

8.1 Power Factor Correction Defined

Power factor (PF) is defined as the ratio of true power in watts (W) to apparent power volt–amperes (VA). A pure resistive load has a PF of 1. But in active loads, such as light-emitting diodes (LED) drivers, the mains supply current passes through a bridge rectifier and then the DC voltage is smoothed using a large-electrolytic capacitor. As the capacitor only charges during the peaks in the AC supply cycle, the current is pulsed and nowhere near sinusoidal. Simple rectified supplies tend to have PFs close to 0.5.

Special measures can be taken to "correct" a poor PF. Power factor correction, or PFC, is a term used with AC mains–powered circuits. Techniques are used to create a good PF, so that the AC current is sinusoidal and in phase with the AC voltage.

There are regulations in many countries that require a good PF. In Europe, regulations demand that residential lighting have a PF of greater than 0.7 and commercial lighting have a PF greater than 0.9. In any case, lighting products with a power rating of greater than 25 W must have a good PF of 0.95 or higher, regardless of the end use. In the United States, there are voluntary guidelines, with the DoE and Energy Star requirements being similar to Europe. The exception is the State of California that asks for a PF greater than 0.9 for *all* power levels of residential and commercial LED lighting.

Distortion of the current waveform reduces the PF, but also gives some harmonic distortion. There is a relationship between PF and total harmonic distortion (THD):

$$PF = \frac{1}{\sqrt{1 + \left(\dfrac{THD}{100}\right)^2}}$$

Electromagnetic compatibility requirements sometimes quote THD, rather than PF. In Europe, applying the limits specified by the European Normalisation, EN-61000-3-2 class C, is mandatory for lighting products.

8.2 Typical PFC Boost Circuit

The most common PFC circuit is a boost converter, where the AC line voltage is boosted to give a constant voltage of about 400 V across an energy storage capacitor. The peak amplitude of the charging current pulses into this capacitor is arranged to be sinusoidal. This is achieved by switching the current on for short but constant periods: with constant on-time switching, as the supply voltage rises and falls, so does the amplitude of the current. A typical PFC circuit using the ST Microelectronics L6562 is shown in Fig. 8.1.

There are a number of controller ICs for PFC. One of the most popular has been the L6562 from ST Microelectronics. Other well-known controllers are HBD853-D

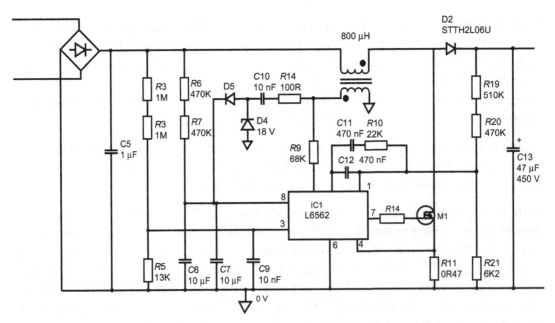

Figure 8.1: Power Factor Correction (PFC) Circuit.

(On-Semiconductor), UCC29019 (Texas Instruments), LT1249 (Linear Technology), and FAN7350 (Infineon).

A simple alternative is to use a fly-back supply, see Section 8.3. It is common to switch the primary current off when a certain current level is reached, but this leads to constant average current. To give a good PF, the primary current should be switched with a constant "on-time," so that the current amplitude rises and falls in phase with the supply voltage. The secondary current will rise and fall at double the AC line frequency and so a large secondary capacitor is required to absorb this ripple, to prevent significant ripple in the output voltage.

Driving a LED from the output of a PF-corrected supply usually requires a simple buck converter, as the voltage source tends to be very high (about 400 V). However, alternative solutions exist; these are the Bi-Bred and the buck–boost–buck (BBB). Both of these topologies can be driven using an HV9931 IC from Microchip, see Sections 8.5 and 8.6. The advantage of using a simple buck regulator from a constant voltage supply is that there will be no low-frequency ripple current through the LEDs, which would be at double the AC mains frequency, so 100 or 120 Hz.

8.3 Boost–Buck Single Switch Circuit

The boost–buck single switch circuit is also known as a fly-back circuit. Fly-back circuits are described in more detail in Chapter 9, but this section is specifically considering their use as direct LED drivers with PFC. In the case of fly-back circuits, constant on-time switching is required in order that the input current follows the sinusoidal input voltage.

The schematic given in Fig. 8.2 was adapted from an On-Semiconductor application note AND9043/D. It shows how an On-Semiconductor NCP1014 can be used to drive an LED load of 35–80 V; typically 10–22 LEDs connected in series. Only a few critical parts are numbered, a more detailed explanation is offered further.

The NCP1014 has a high-voltage switch between pin 3 and pin 4 (0 V). The switch turns on for a constant period and the inductor is charged. Then when the switch turns off, current continues to flow through $L1$ via diode D1, charging the output capacitors $C9$ and depleting the inductor's energy. Once the voltage across $C9$ is sufficiently high, current will flow through the LEDs. The current flow causes a voltage drop across $R10$; this voltage is then used to provide feedback to the control IC. Feedback is via a photocoupler; the voltage across $R10$ feeds a potential divider $R11$ and $R14$ to drive a small current through the photocoupler's LED. When the corresponding phototransistor turns on, pin 2 of the NCP1014 is pulled low and the switching is disabled.

In addition, if the LEDs become open circuit, some protection is needed to prevent damage of $C9$ (which could explode) and burning of components. If the output voltage rises above

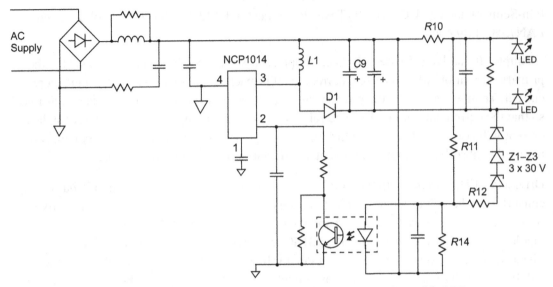

Figure 8.2: Buck–Boost Light-Emitting Diode (LED) Driver with PFC.

the nominal 90 V of the protection circuit Zener diodes, Z1–Z3, current flows into the photocoupler's LED. The switching is disabled, as before.

The circuit shown in Fig. 8.2 will produce light with almost 100% low-frequency ripple (100 or 120 Hz) because there is very little energy storage in capacitor C9. When the AC mains voltage is close to the zero crossing point, there will be no light from the LEDs. This prevents such driver circuits from being used in applications, such as traffic lights, where the high level of low-frequency ripple would interfere with the associated safety camera. This is because the frame scan rate of a camera is close to 100 Hz and the resultant image would show a flicker effect.

8.4 Boost–Linear Regulator Circuit

One technique used in many high-power LED lamps, such as streetlights, is to boost the voltage to about 400 V using a PFC stage as described at the beginning of this chapter. The current is limited, but not controlled. To control the current and drive the LEDs, simple buck switching regulators can be used to provide a constant current source.

An alternative, to be described in this section, is to boost the voltage and then use a linear regulator to control the LED current. To achieve good efficiency, the system has to be designed carefully. As boosting universal AC mains voltages results in typically 450 V being produced, the LED forward voltage drop must also be about 440 V, to leave 10 V across the linear regulator. To achieve such a high-forward voltage drop, a lot of LEDs must be connected in series (typically 120–140 LEDs are needed).

With a high number of LEDs, the forward voltage tolerance will also be significant. If the boost stage voltage was 450 V and the total voltage drop of the LED string was of a similar voltage, due to a batch of LEDs with particularly high-forward voltage drop (V_f), the LEDs would be dim or would not light. To overcome this, the output voltage of the boost stage needs to be adjusted automatically so that the required LED current is reached.

Conversely, if the total forward voltage of the LED string were lower than expected, the voltage drop across the linear regulator would be higher and hence, the power dissipation could be too high. Again, an output voltage control of the boost regulator, so that the voltage drop across the linear regulator is as low as possible, will solve this problem.

One such circuit to perform the functions described above is the HV9805 from Microchip. A typical schematic using the HV9805 is given in Fig. 8.3.

The boost section of the circuit uses a 600-V MOSFET, M1, in a cascade connection. The gate of M1 is connected to an 18-V supply, comprising R1, R2, Z1, and C4, powered from the rectified mains. There is an internal MOSFET connected between pin 10 and pin 1 (GND), which drives the source of M1 to ground and turns it on. The internal MOSFET thus never sees more than 18 V across its drain and source, allowing low-voltage (and low-cost) silicon processes to be used.

The MOSFET switching is controlled by monitoring the voltage across R8, which returns the ground connection to the bridge rectifier. The CSH and CSL pins are differential inputs used to sense the current. After the MOSFET M1 turns off, there will be some ringing at its drain. The circuit needs to detect when the ringing has stopped, before it starts the next switching cycle, otherwise the PF could be affected. However, when the AC input voltage is low, near to the zero crossing point, the ringing is difficult to detect. The addition of C9 (with a small damping resistor R18 in series) increases modulation of the sense voltage, to improve detection at these times.

The linear regulator at the output uses an external 300-V MOSFET M2. The voltage rating can be lower than the high-voltage rail because most of the output voltage drops across the LEDs. Resistor R13 is the output current sense resistor. The voltage across M2 is monitored by the potential divider R6 and R7 and input to pin 4 (HVS). The voltage on pin 4 is proportional to the headroom voltage on the linear regulator, and is adjusted by controlling the output of the boost stage.

8.5 Bi-Bred

The Bi-Bred is very similar to the Cuk boost–buck that was described in the previous chapter (see Fig. 8.4).

The main difference between the Cuk and the Bi-Bred is that, in a Bi-Bred, the input inductor is in discontinuous conduction mode (DCM) and operation of the output stage in continuous

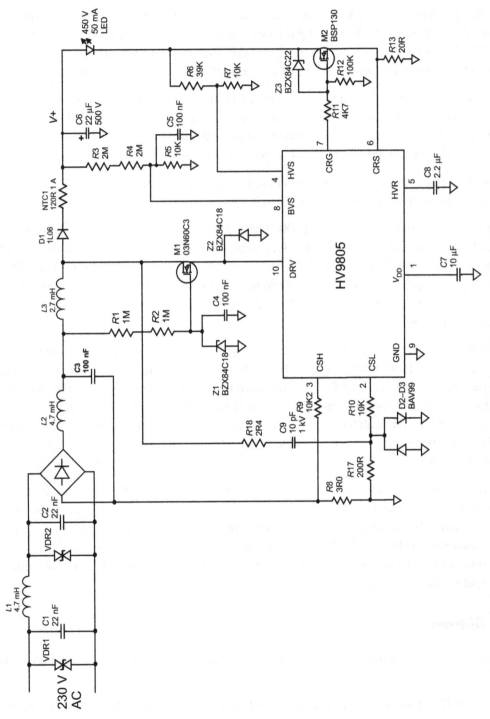

Figure 8.3: Boost–Linear Regulator Using HV9805.

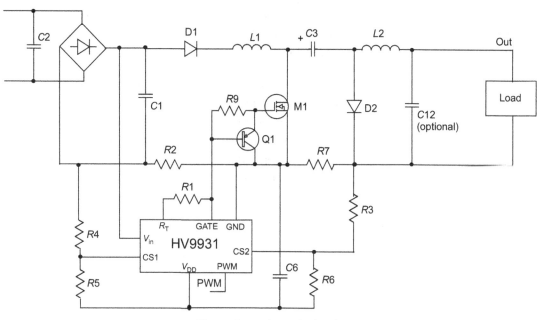

Figure 8.4: Bi-Bred Circuit.
PWM, Pulse width modulation.

conduction mode (CCM). The energy stored in each inductor is proportional to the inductance value. This means that in the design, the input inductor $L1$ must have a small enough energy stored to ensure that conduction stops before the end of each cycle. Consequently, the input inductor value must be relatively small. The output inductor $L2$ must have large enough energy stored (large-inductance value), so that the current only falls to about 85% of its nominal value at the end of each switching cycle.

When power is first applied, MOSFET M1 is off and waiting for the first clock signal to trigger the gate drive pulse. At this time, the storage capacitor $C3$ immediately begins to charge from the supply voltage through D1 and $L1$. The voltage across $C3$ will not rise very high though because, when the MOSFET M1 switches on, the charging current is redirected to the 0-V rail. With M1 conducting, current continues to rise in amplitude through the inductor $L1$ until the voltage drop across $R2$ is sufficient for the input current feedback comparator inside the HV9931 to trigger, which turns M1 off. Now the input circuit acts like a boost converter because the current through $L1$ cannot change immediately and it charges $C3$ to a high voltage.

The next time that M1 switches on, energy stored in $C3$ is used to drive current through the LED load. The current rises in inductor $L2$ and the load until the voltage drop across resistor $R7$ is sufficient to trip the HV9931's output current sense comparator and turn M1 off again. The current flow through $L2$ passes through D2 to keep current flowing in the LED load.

Notice that the current sense resistor is not in this path because the current level measurement is not required until the MOSFET turns on again. Having the current sense resistor outside the flywheel diode path minimizes power loss.

The output of the Bi-Bred is configured as a buck stage. Energy is supplied from a bulk storage capacitor, $C3$. This capacitor has a sufficiently large capacitance value to provide a more or less constant supply voltage over an AC line cycle period. A constant capacitor voltage supplying power to the buck stage means a constant switch duty cycle when it is driving the LED load. The Bi-Bred draws an approximately sinusoidal AC line input current when driven from a switch operating at constant duty cycle, hence a large capacitance value for $C3$ helps to produce a good PF.

The duty cycle of the switching is given by $\dfrac{V_o}{V_i} = \dfrac{D}{1-D}$.

Or put another way, $D = \dfrac{V_o}{V_i + V_o}$. So if $V_{in} = 350$ V and $V_o = 3.5$ V (a typical white LED),

$D = \dfrac{3.5}{350 + 3.5} = \dfrac{3.5}{353.5} = 0.99\%$. Such a small duty cycle can be difficult to switch properly, as we have seen for a buck converter. This means that a Bi-Bred is not really suitable for driving short LED strings.

8.6 Buck–Boost–Buck

The buck-boost-buck (which we will refer to as BBB) is a proprietary circuit, patented by Supertex (now Microchip), and is illustrated in Fig. 8.5. It resembles the Bi-Bred in some respects, except for two current-steering diodes, D1 and D4.

Like the Bi-Bred, the input inductor is in DCM and operation of the output stage in CCM. The energy stored in each inductor is proportional to the inductance value. This means that in the design, the input inductor $L1$ must have a small enough energy stored to ensure that conduction stops before the end of each cycle. Hence the input inductor value must be relatively small. The output inductor $L2$ must have large enough energy stored (large-inductance value) so that the current only falls to about 85% of its nominal value at the end of each switching cycle.

When power is first applied, MOSFET M1 is off and waiting for the first clock signal to trigger the gate drive pulse. At this time, the storage capacitor $C3$ is not charged. With M1 conducting, current begins to rise in amplitude through the inductor $L1$ until the voltage drop across $R2$ is sufficient to trip the HV9931's input current sense comparator, which turns M1 off. Now the input circuit is in flywheel mode because the current through $L1$ cannot change immediately and it charges $C3$ to a moderately high voltage. The voltage across $C3$ is typically midway between the input and output voltage levels.

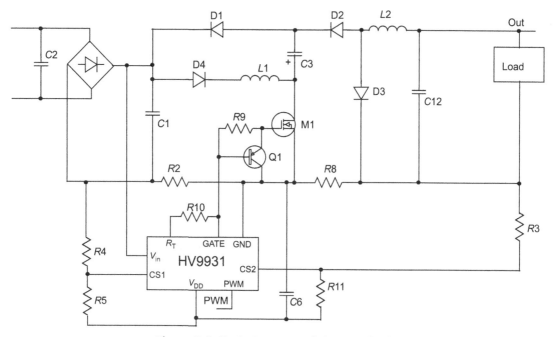

Figure 8.5: Buck–Boost–Buck (BBB) Circuit.

The energy in *C3* is used to drive current through D2, *L2*, and the LED load the next time that M1 switches on. The current rises in inductor *L2* and the load until the voltage drop across resistor *R8* is sufficient to trip the output current sense comparator and turn M1 off again. The current flow through *L2* passes through D2 to keep current flowing in the LED load. Notice that, like in the Bi-Bred, the current sense resistor is not in this path because the current level measurement is not required until the MOSFET turns on again; this minimizes power loss.

The output of the BBB is the buck stage. Energy is supplied from a bulk storage capacitor *C3*, with sufficiently large capacitance to provide a more or less constant supply voltage over an AC line cycle period. A constant capacitor voltage across *C3*, which supplies power to the buck stage, means a constant switch duty cycle when it is driving the LED load. The BBB draws a more or less sinusoidal AC line input current when driven from a switch operating at constant duty cycle. Hence, like the Bi-Bred case, a large-capacitance value for *C3* helps to produce a good PF.

In practice there is a limit to the value of *C3* because of physical size, particularly if plastic film capacitors are used. This means that there will be some voltage ripple across *C3*, at a frequency double that of the AC line (i.e., 100 Hz when driven from a 50-Hz supply, or 120 Hz when driven from a 60-Hz supply). The effect of this ripple voltage is to generate secondary harmonic signals in the input current, which reduces the PF. By adding a simple circuit, the second harmonic can be reduced; the MOSFET off-time is modulated by the ripple voltage and this acts like a negative feedback to reduce the second harmonic.

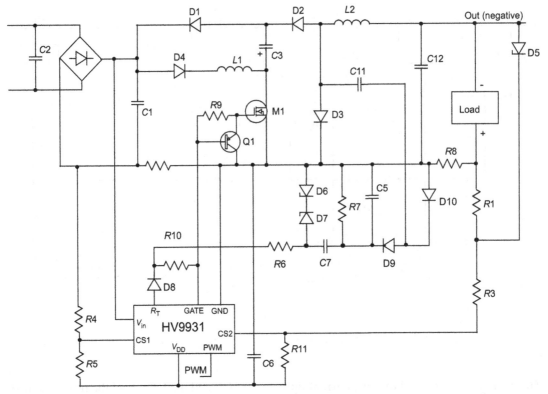

Figure 8.6: BBB With Harmonic Reduction.

The additional circuits are usually used only if plastic film capacitors or small-value electrolytic capacitors are used for $C3$ and are shown in Fig. 8.6.

When MOSFET M1 is conducting, there is a current path from $C3$ through D10 and D2 to charge $C11$ (with a negative voltage at the anode of D3 and positive at the anode of D9). When MOSFET M1 turns off, so that D3 is conducting the LED current, the voltage across $C11$ can forward bias D9 and charge capacitor $C5$. Between each switching cycle, resistor $R7$ discharges capacitor $C5$. Thus the ripple voltage across $C3$ will modulate the average voltage across $C5$. Capacitor $C7$ acts as a DC block, to allow just the modulation across $C5$, rather than any DC level, to vary the MOSFET off-time. As the voltage across $C5$ rises and falls, current through $R6$ rises and falls, thus shortening or lengthening the off-time in proportion to the AC mains voltage.

The duty cycle of the switching in a BBB converter is given by $\dfrac{V_o}{V_i} = \dfrac{D^2}{1-D}$.

Or put another way, $D = \dfrac{-V_o \pm \sqrt{V_o^2 + 4 \cdot V_i \cdot V_o}}{2 \cdot V_i}$. So if $V_{in} = 350$ V and $V_o = 3.5$ V (a typical

white LED), $D = \dfrac{-3.5 \pm 70}{700} = \dfrac{66.5}{700} = 9.5\%$. This is a considerably greater duty cycle than either the Bi-Bred or the buck converter with a similar low-voltage load. This means that the BBB converter is most suitable for driving short LED strings.

8.6.1 Buck–Boost–Buck Design Equations

The HV9931 is a peak current control IC that is designed for controlling the nonisolated single-stage PFC converter (BBB) described earlier. A typical application circuit of the HV9931 was shown in Fig. 8.5, but is redrawn with different component annotation to help with the further description, in Fig. 8.7.

The HV9931 has a built-in high-voltage regulator circuit producing 7.5 V ± 5% at V_{DD}, hence making the circuit relatively simple because we do not have to consider separately powering the integrated circuit.

As soon the start-up threshold is reached at V_{DD}, an internal oscillator circuit is enabled. The oscillator circuit is actually three functions: a timer, a latch, and a latch-reset circuit. The timer is a current mirror that charges an internal capacitor; when the threshold of 0.63 V_{DD} is reached (after $R_T \times C$ seconds, where C is the internal capacitor) the latch is triggered. At this time, the timer capacitor is discharged ready for the next cycle. The output signal of the latch turns on the GATE output and switches on the power MOSFET Q1. The oscillator circuit also includes two comparators; if either of these is triggered, the latch is reset and the MOSFET Q1 turns off.

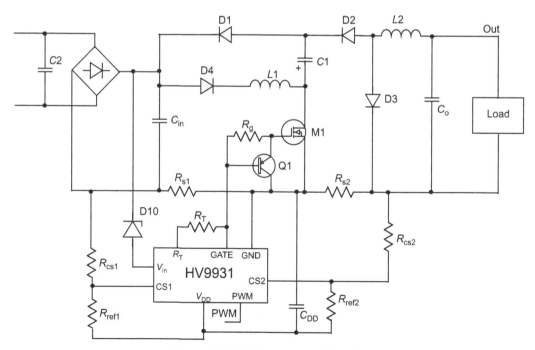

Figure 8.7: BBB-Annotated Circuit.

The timer uses a current mirror, with one-half connected to the R_T pin. The current mirror needs a small current to flow to ground (0 V) from the R_T pin, which causes the other half of the mirror to charge an internal timing capacitor. This means that the timer can be programmed with a single resistor connected to R_T for either constant switching frequency or fixed off-time operation.

In the constant frequency mode, a constant timing current is required, so the timing resistor must be connected between the R_T pin and 0 V. In the fixed off-time mode, the timing resistor must be connected between the GATE and R_T pins; current flow (and hence timing) only begins when the GATE pin is at 0 V. Thus the timer will set the latch and turn on the MOSFET after a programmed time period following the turn-off of the GATE output.

The fixed off-time operating mode is preferred because it: (1) reduces the voltage stress at $C1$, (2) improves input AC ripple rejection, and (3) inherently introduces frequency jitter that can help reduce the size of the input EMI filter required. Hence we will use the fixed off-time mode for the following design.

Connecting the resistor from R_T to GATE programs constant off-time:

$$T_{\text{off}} = \alpha \cdot R_T + \tau_0$$

Where $\alpha = 40$ pF, $\tau_0 = 880$ ns.

In terms of R_T, this transposes to:

$$R_T = \frac{T_{\text{off}} - \tau_0}{\alpha} = \frac{T_{\text{off}} - 880\,\text{ns}}{40 \cdot 10^{-12}}$$

The input and output feedback comparators are used as current sense inputs, for programming peak currents in $L1$ and $L2$. Both comparators use the ground potential (GND) as a reference and can be used to monitor voltage signals of negative polarity with respect to GND. A blanking delay of 215 ns is added to prevent false tripping the comparators due to the circuit parasitic capacitance. The currents i_{L1} and i_{L2} that trip the comparators can be computed as:

$$i_{\text{Lpk}} = \frac{V_{\text{ref}} \cdot R_{\text{cs}}}{R_{\text{ref}} \cdot R_{\text{s}}}$$

Where V_{ref} in this equation is an external reference voltage.

We will use $V_{\text{ref}} = V_{\text{DD}}$ as an example. When either of the feedback comparators detects negative input voltage at its current sense (cs) input, the latch resets, the GATE output becomes low voltage.

We can easily calculate the current in L2, as the output current is more or less constant with some ripple. It is assumed to operate in CCM,

$$i_{L2pk} = i_{L2} + 0.5 \cdot \Delta i_{L2}$$

Where i_{L2} is the average current and Δi_{L2} is the peak-to-peak current ripple in L2. Thus the constant peak current control used in the HV9931 introduces a peak-to-average error $0.5 \times \Delta i_{L2}$ that needs to be accounted for when programming the resistor divider R_{ref2}/R_{cs2}. Fortunately, this error is nearly constant for any input voltage at fixed T_{off} and it is relatively small compared to i_{L2} (15% typical). The ripple will have a minimal effect on the overall regulation of the output current. The error is however a function of the output voltage variation and the inductance value tolerances of L2.

8.6.1.1 Choice of L1

We need to design the input buck–boost stage to operate in DCM to give a good PF. Operating in DCM at any given line and load condition will ensure low distortion of the input current and stability of the control loop. Therefore, let us assume that the current in L1 becomes critically continuous at full load and at some minimum operating AC line voltage $V_{AC\,min}$. This boundary conduction mode (BCM) condition should normally occur at the peak of each half-wave of the input AC current. If we assume a unity PF (PF = 1), this boundary condition will then coincide with the peak input voltage $V_{AC\,min}$. As both converter stages are in CCM at this point, the ratio between the output and the input voltage can be expressed as:

$$\frac{V_o}{V_{AC\,min} \cdot \sqrt{2}} = \frac{D_{max} \cdot \eta_1}{1 - D_{max}} \cdot D_{max} \cdot \eta_2 = \frac{D_{max}^2 \cdot \eta}{1 - D_{max}}$$

Where η_1 and η_2 are the corresponding efficiencies of the input buck–boost stage and the output buck stage. The overall converter efficiency equals $\eta = \eta_1 \times \eta_2$. The duty ratio D of the switch M1 is the greatest at this condition. (Duty ratio is defined as $D = T_{on}/T_s$, where T_{on} is the on-time of M1, and T_s is the switching period.)

The input AC line current is a ratio of the output power to the input voltage and efficiency:

$$i_{AC} = \frac{V_o \cdot i_o}{V_{AC} \cdot \eta}$$

On the other hand, the peak input current must be equal to the (averaged) peak current through L1:

$$i_{AC} \cdot \sqrt{2} = \frac{D}{2} \cdot i_{L1pk}$$

Where the peak current in $L1$ (i_{L1pk}) is dependent on the maximum voltage across $L1$, the inductance and the on-time:

$$i_{L1pk} = \frac{V_{AC} \cdot \sqrt{2} \cdot T_{on}}{L1}$$

As we are only considering the constant off-time case, let us express this in terms of $T_{off} = T_{on} \times (1-D)/D$.

$$i_{L1pk} = \frac{V_{AC} \cdot \sqrt{2} \cdot T_{off}}{L1} \cdot \frac{D}{1-D}$$

Combining the equations (using the mean value of $V_{AC\,min}$) and solving for the inductance value gives:

$$L1 = \frac{V_{AC\,min} \cdot \sqrt{2} \cdot T_{off}}{4 \cdot i_o}$$

Note, that the critical inductance $L1$ in this equation corresponds to the boundary conduction at $V_{AC\,min}$. The value used in practice should be lower than this, to allow for component tolerances and parasitic elements. The inductor value has some tolerance, perhaps 10%, but tolerance in all the other components and switching off-time should be considered too. If the inductance goes above the critical value, CCM will occur and the input waveform will be severely distorted. For these reasons, the value of $L1$ should be reduced to 63% of its maximum allowed value:

$$L1 = \frac{0.63 \cdot V_{ACmin} \cdot \sqrt{2} \cdot T_{off}}{4 \cdot i_o}$$

The designer must be careful when considering standard inductors for $L1$ or designing a custom one. As $L1$ conducts discontinuous current, magnetic flux excursion in the core material can cause heating. Hence the design of $L1$ is limited by the power dissipation in the magnetic core material rather than by the saturation current of the inductor selected.

8.6.1.2 Choosing C1

Capacitor $C1$ is usually an electrolytic capacitor because of the required high-capacitance value and high-voltage rating. This is sometimes a cause for concern about reliability. However, the lifetime rating of electrolytic capacitors has improved in recent years and lifetime ratings of 10,000 h at 105°C are available from multiple sources. Considering that the lifetime doubles for every 10°C drop in temperature, this equates to 40,000 h at 85°C. So at reasonable temperatures, the lifetime of the electrolytic and the LED are comparable.

Capacitor $C1$ is the main energy storage element and its capacitance value significantly affects the distortion of input current. The minimum capacitance value of $C1$ is determined by the input harmonics limits required for a specific application. Lighting products are sold in large quantities, and thus these high-volume products can potentially have a high impact on the low-voltage public supply system. The European EN 61000-3-2 class C limits are applicable to lighting products and are comparable to the limits imposed by ANSI C82.77 standards in the US market. These restrict THD to approximately 33%. Both, the class C and ANSI standards limit the third harmonic current of lighting products to ~30%. The regulations for LED-based traffic signal heads are generally stricter and require THD to be less than 20% (ITE VTCSH Part 2).

The main component of the AC ripple voltage across $C1$ is the second AC line harmonic. This ripple causes modulation of the duty cycle, $D(t)$, given by the following equation:

$$D(t) = \frac{V_o}{\eta_2 \cdot V_C(t)}$$

Where $V_C(t)$ is voltage across $C1$ at any time t.

Combining the equations developed in Section 8.6.1.1, we have an equation for the input AC current:

$$i_{AC} = \frac{D}{2\sqrt{2}} \cdot \frac{V_{AC} \cdot \sqrt{2} \cdot T_{off}}{L1} \cdot \frac{D}{1-D}$$

In terms of a sinusoidal waveform, this becomes:

$$i_{AC}(t) = \frac{V_{AC} \cdot \sqrt{2} \cdot T_{off}}{2 \cdot L1} \cdot \frac{D^2(t)}{1-D(t)} \cdot \sin(2\pi \cdot f_{AC} \cdot t)$$

Where f_{AC} is the AC line frequency (50 or 60 Hz).

Assume that a small second harmonic ripple voltage V_C exists across $C1$, i.e. 100 or 120 Hz, so that the voltage at $C1$ can be written as:

$$V_C(t) = V_C - v_C \cdot \sin(4\pi \cdot f_{AC} \cdot t)$$

Where $i_c/V_C \ll 1$. By substituting across these last three equations we will produce a displaced fundamental term and a third harmonic term in the AC line current. It can be shown from the resulting equation that the third harmonic distortion of the input AC line current for a given relative second harmonic ripple $K_C = i_c/V_C \ll 1$ is:

$$K_3 = \frac{\Delta i_3}{i_{AC}} \approx 0.5 \cdot K_C \cdot \frac{2-D}{1-D}$$

Thus every 1% of second harmonic ripple at $C1$ will generate at least 1% of third harmonic component in the AC line current even when the duty cycle is small.

Now we can find the capacitance value of $C1$ needed to limit the third harmonic distortion to some given constant K_3. Equations used to find $L1$ can now be used to solve for the duty cycle D at any V_{AC} within the operating range.

$$D = \frac{V_o \cdot i_o \cdot L1}{V_{AC}^2 \cdot T_{off} \cdot \eta} \cdot \left(\sqrt{1 + \frac{2 \cdot V_{AC}^2 \cdot T_{off} \cdot \eta}{V_o \cdot i_o \cdot L1}} - 1 \right)$$

This is a bit unwieldy, so let us introduce a parameter δ as follows:

$$\delta = \frac{2 \cdot V_{AC}^2 \cdot T_{off} \cdot \eta}{V_o \cdot i_o \cdot L1}$$

Now the equation for the duty cycle looks much simpler:

$$D = \frac{2 \cdot \left(\sqrt{1+\delta} - 1 \right)}{\delta}$$

We can rewrite the equation for the third harmonic ripple K_3 as:

$$K_3 = K_C \cdot \frac{1}{1 - \frac{1}{\sqrt{1+\delta}}}$$

Recalling that $D = V_o/(\eta_2 \times V_C)$ and using $D = 2 \left(\sqrt{(1+\delta)} - 1\right)/\delta$ from above, we can determine the voltage at $C1$ for a given V_{AC}:

$$V_C = \frac{V_o}{2 \cdot \eta_2} \cdot \left(1 + \sqrt{1+\delta} \right)$$

We have assumed that $K_C = i_C/V_C \ll 1$. This condition is met if the value of capacitor $C1$ large enough, so that the AC ripple voltage across $C1$ is low. Therefore, $C1$ decouples the bulk of the AC ripple current at the output of the input converter stage. Averaged over a switching cycle, this current can be written as:

$$i_2(t) = \frac{V_{AC}(t) \cdot i_{AC}(t) \cdot \eta_1}{V_C}$$

The AC line current $i_{AC}(t)$ has been given by a previous equation. Based on the assumptions made earlier, the AC component of $i_2(t)$ contains second harmonic current only. This AC current in $C1$ can be expressed as:

$$i_C(t) = -\frac{V_{AC}^2 \cdot \eta_1 \cdot T_{off}}{2 \cdot L1 \cdot V_C} \cdot \frac{D^2}{1-D} \cdot \cos(4\pi \cdot f_{AC} \cdot t)$$

Substituting D and V_C (using δ from the condensed equations earlier) gives:

$$i_C(t) = -\frac{2 \cdot i_o}{1+\sqrt{1+\delta}} \cdot \cos(4\pi \cdot f_{AC} \cdot t)$$

Relative ripple voltage at $C1$ can be calculated as $K_C = i_{Cpk} \times Z_C/V_C$, where i_{Cpk} is the amplitude of $i_C(t)$ and $Z_C = 1/(4\pi \times f_{AC} \times C1)$ is the impedance of $C1$ at $2 \times f_{AC}$. Substituting V_C we obtain:

$$K_C = \frac{1}{\left(1+\sqrt{1+\delta}\right)^2} \cdot \frac{\eta_2 \cdot i_o}{\pi \cdot f_{AC} \cdot C1 \cdot V_o}$$

Solving this equation for $C1$ and substituting for K_C we get:

$$C_1 = \frac{1}{\delta \cdot \left(1 + \frac{1}{\sqrt{1+\delta}}\right)} \cdot \frac{\eta_2 \cdot i_o}{\pi \cdot f_{AC} \cdot K_3 \cdot V_o}$$

The RMS value of the switching current in $C1$ can be calculated using the following equation:

$$i_{C\,SW} = i_o \cdot \sqrt{\frac{64}{9 \cdot \pi \cdot \eta \cdot \eta_1} \cdot \frac{V_o}{V_{AC} \cdot \sqrt{2}} + D}$$

The RMS value of the second AC line harmonic is:

$$i_{C\,line} = \frac{\sqrt{2} \cdot i_o}{1+\sqrt{1+\delta}}$$

8.6.1.3 Calculating L2

Calculating the value of the output filter inductor $L2$ is simple. The designer must decide on the allowable amount of switching ripple current in $L2$. Then:

$$L2 = \frac{V_o \cdot T_{off}}{\Delta i_{L2} \cdot \eta_2}$$

Where Δi_{L2} is the peak-to-peak current ripple in $L2$. Larger values of $L2$ will produce smaller ripple Δi_{L2}, and therefore smaller peak-to-average error in the output current control loop. However, low-amplitude feedback signals would also make the output current sense comparator more susceptible to noise. It is a good practice to design L2 for $\Delta i_{L2} = 20$–30%. The reason for having moderate inductor current ripple is to allow a sufficiently high-feedback signal for the current sense comparator. An output capacitor connected across the LED load can be added, to reduce the output ripple current further, if needed.

Unlike the design of input inductor L1, the design of L2 is typically limited by the saturation flux of its magnetic material. However, power dissipation due to the core loss may also need to be considered. The saturation current rating of the inductor must satisfy:

$$i_{\text{sat}} > i_\text{o} + 0.5 \cdot \Delta i_{L2}$$

8.6.1.4 Power semiconductor components

Now we can calculate the voltage and the current ratings of the MOSFET M1 and the rectifiers D1–D4.

8.6.1.4.1 MOSFET M1

The current in M1 is composed from the currents in the inductors L1 and L2. Hence, the RMS current in M1 can be computed as:

$$i_{\text{D M1}} = \sqrt{\frac{D_{\max} \cdot i_{L1\text{pk}}^2}{6} + D_{\max} \cdot i_\text{o}^2}$$

Where $i_{L1\text{pk}}$ and $D_{\max}$ are calculated at $V_{\text{AC min}}$. We can disregard the ripple current in L2, as we want the average current. The drain voltage rating of M1 can be determined as:

$$V_{\text{DSM1}} = V_{\text{AC max}} \cdot \sqrt{2} + V_{\text{C max}} \cdot \left(1 + K_\text{C}\right)$$

Where $V_{\text{C max}}$ and K_C are calculated at $V_{\text{AC max}}$.

For the power MOSFET M1, it is very important to find a good balance between the total gate charge Q_g and the on resistance RDS (ON). Using a MOSFET with low RDS (ON) will not necessarily achieve greater efficiency. The gate drive current from the HV9931 is limited, so a large-gate charge will slow down the switching time and thus increase switching losses.

In addition to the slow turn-on speed generating higher-switching power loss, MOSFETs with high Q_g will require more current from the HV9931's internal voltage regulator. When a MOSFET is being switched, the average current required by the gate driver is given by: $i_\text{g} = Q_\text{g} \times f_{\text{sw}}$. So, if we have a MOSFET with 30-nC gate charge being switched at 50 kHz, the average gate driver current is 1.5 mA. A small-additional current is required by the

HV9931 internal bias circuits, so we could have 2 mA drawn from the internal high-voltage regulator, while dropping 300 V or so. The HV9931 gate driving capability is therefore also limited by the package power dissipation. Being a linear regulator, the power dissipated will be $i_{in} \times (V_{in} - V_{DD})$. Thus nonoptimal selection of M1 may cause the HV9931 to overheat.

8.6.1.4.2 Diodes D1–D4

The highest currents in D1–D4 averaged over the AC line cycle can be calculated as:

$$i_{D1} = \frac{4 \cdot \sqrt{2}}{\pi} \cdot \frac{i_o}{\eta_1 \cdot \left(1 + \sqrt{1 + \delta_{min}}\right)}$$

$$i_{D2} = D_{max} \cdot i_o = \frac{2 \cdot \left(\sqrt{1 + \delta_{min}} - 1\right)}{\delta_{min}} \cdot i_o$$

$$i_{D3} = \left(1 - D_{min}\right) \cdot i_o = \frac{\left(\sqrt{1 + \delta_{max}} - 1\right)^2}{\delta_{max}} \cdot i_o$$

$$i_{D4} = \frac{4 \cdot \sqrt{2}}{\pi} \cdot \left(\frac{2 \cdot \sqrt{2}}{\delta_{min}} + \frac{1}{\eta_1 \cdot \left(1 + \sqrt{1 + \delta_{min}}\right)}\right) \cdot i_o$$

Where δ_{max} and δ_{min} are calculated from our definition of δ given in Section 8.6.1.2, at $V_{AC\,max}$ and $V_{AC\,min}$ correspondingly. Peak currents in D1 and D4 equal to i_{L1pk} are calculated using the value of δ at $V_{AC\,min}$. Peak currents in D2 and D3 are computed as $i_o + 0.5 \times \Delta i_{L2}$.

The voltage ratings for D1–D3 are given as:

$$V_{RD1} = V_{AC\,max} \cdot \sqrt{2} + V_C \cdot \left(1 + K_C\right)$$
$$V_{RD2} = V_{AC\,max} \cdot \sqrt{2}$$
$$V_{RD3} = V_C \cdot \left(1 + K_C\right)$$

Using ultrafast recovery rectifiers for D2 and D3 is essential for good efficiency of the LED driver. Both diodes operate at high current and are subjected to fast transitions and high-reverse voltage.

The required reverse voltage rating of D4 depends on several factors. During the dead time, there could be a significant postconduction resonance. The LC tank is formed by L1 and the parasitic capacitance of D1, D4, and M1. The resonant period can be estimated as:

$$T_R = 2 \cdot \pi \cdot \sqrt{\frac{L1 \cdot C_{JD4} \cdot \left(C_{OSS} + C_{JD1}\right)}{C_{OSS} + C_{JD1} + C_{JD4}}}$$

Where C_{OSS} is the output capacitance of M1, and C_{JD1} and C_{JD4} are reverse-biased junction capacitances of D1 and D4 correspondingly.

Due to a finite reverse recovery time of D1, the input inductor $L1$ develops a certain reverse current in the beginning of the dead time. As $L1$ runs in DCM, the reverse recovery time of D1 has negligible effect from the overall power efficiency point of view. However, even a small reverse current in $L1$ can cause a very high-voltage spike across D4, when both diodes stop conducting. Thus, ultrafast recovery diode is recommended for D1.

As $C_{JD4} << C_{OSS}$ typically, the postconduction oscillation occurs mainly across D4. The drain voltage of M1 will remain almost unchanged throughout the dead time. Besides causing the high-voltage stress across D4, this oscillation may affect the EMI performance of the circuit. Thus, adding an RC snubber circuit across D4 is recommended. If the snubber capacitance value is greater than $(C_{OSS} + C_{JD1})$, the reverse voltage rating of D4 can be reduced significantly.

The snubber capacitor can be selected as (approximately)

$$C_D = 10 \cdot C_{OSS}$$

And the snubber resistor

$$R_D = \frac{0.33}{C_D \cdot 2 \cdot \pi \cdot f_s}$$

The snubber resistor should be rated at 0.5 or even 1 W, and 500 V breakdown. The minimum voltage rating for D4 will then be:

$$V_{RD4} = V_{ACmax} \cdot \sqrt{2}$$

Thus in theory a fast 400-V diode can be used for D4 in a universal 90–260 V AC LED driver. In practice a fast diode rated at 600 V is normally used.

8.6.1.5 Output open circuit and input undervoltage protection

The BBB circuit is a constant output current source. Hence it can generate destructive voltage at its output in the case of an open circuit condition. A simple circuit shown in Fig. 8.8 protects the HV9931 LED driver from the output overvoltage.

Zener voltage of D12 greater than the maximum output voltage must be selected. Resistor R_{ov} is typically 100–200 Ω. However, it will affect the output current divider ratio and needs to be included in the calculations by replacing R_{cs2} by $(R_{cs2} + R_{ov})$.

Note, that the open circuit condition can create an overvoltage across C1. This voltage stress can be limited by connecting a Zener diode or TVS across C1 limiting the voltage to some

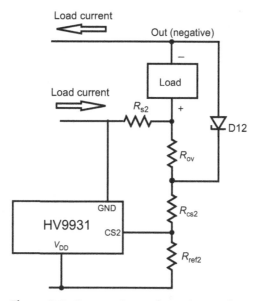

Figure 8.8: Output Overvoltage Protection.

acceptable level greater than $V_{C\,\text{max}}$. The power dissipation in this voltage clamp device is usually small, as the HV9931 operates at minimum duty cycle during the open circuit condition.

The HV9931 inherently protects the LED driver from an input undervoltage condition by limiting the input current. However, increased input current may generate excessive power dissipation in $L1$, D4, M1, and R_{cs1}. Additional protection is recommended by connecting a Zener diode in series with the V_{in} pin of the HV9931, as shown in Fig. 8.9. When the input voltage drops so that the Zener diode stops conducting, the HV9931 no longer has power and turns off.

In addition, an improved input undervoltage protection circuit is shown in Fig. 8.9, which can achieve better performance compared to the simple fixed input current limiting. The reference for the CS1 comparator is derived from the input rectified AC waveform. The voltage divider ratio of $R1:R_{ref1}$ is programmed such that the Zener diode Z_{ref1} clamps the divider voltage at any input greater than $V_{AC\,\text{min}}$, that is:

$$R_1 = R_{ref1} \cdot \frac{V_{AC\,\text{min}} \cdot \sqrt{2} - V_{Z_{ref1}}}{V_{Z_{ref1}}}$$

$V_{ref} = V_{Z_{ref1}}$ should be used in the earlier equations to find the peak current limit for $L1$, within the normal operating input AC voltage range.

When the input voltage falls below $V_{AC\,\text{min}}$, the reference voltage will reduce too, preventing the inductor $L1$ from entering CCM. Operating $L1$ in CCM can cause undesirable LED

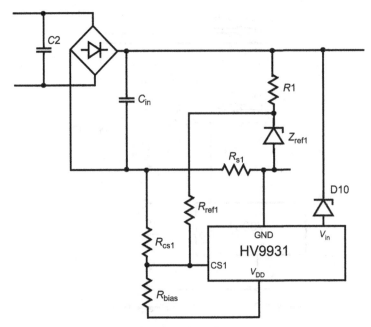

Figure 8.9: Input Undervoltage Protection.

flickering, audible noise, and excessive heat dissipation due to the loop oscillation. R_{bias} creates a positive offset voltage to maintain the reference above 0 V during the AC line voltage zero crossings and thus prevents interruptions of the output current.

8.6.1.6 Surge immunity and EMI considerations

High-voltage surges occur on the AC power mains as a result of switching operations in the power grid and from nearby lightning strikes. LED lighting and signal equipment may be subjected to surge immunity compliance testing in accordance with various standards (EN61000-4-5, NEMA TS-2 2.1.8, etc.) to ensure its continued reliable operation if subjected to realistic levels of surge voltages. The BBB LED driver circuit relies mainly on the transient suppressors (MOV and TVS) to protect it from the input AC line surge. There is little capacitance available at the AC input of the LED driver to absorb high-surge energy, although the EMI filter inductors do slow down any fast transient edges. Thus a transient suppressor needs to be connected across the AC input terminals.

As with all switching converters, selection of the input filter is critical to obtaining good EMI. The HV9931 solution using fixed off-time switching provides an inherent advantage of the frequency dither (spread spectrum). Hence less filtering may be needed, resulting in a smaller EMI filter.

Some important guidelines for PCB design must be followed for optimal EMI performance of the BBB power converter. The area of the fast-switching current loops must be minimized. The first loop including of M1, $C1$, D2, and D3 can significantly degrade the overall EMI performance due to the reverse recovery current in D3. Using a very fast diode is recommended for D3. The second loop consists of C_{in}, D1, C1, M1, and R_{s1}. As the input buck–boost stage runs in DCM, the reverse recovery current in D1 is insignificant. However, charging its junction capacitance can generate fast-current transients. The large physical dimensions of C1 can complicate optimal routing of these loops.

Optimal routing of the HV9931 gate drive output loop can be important for EMI performance, as well as for preventing destructive oscillations of the M1 gate voltage. Adding a 100R resistor in series with the gate will help to dampen any oscillations caused by parasitic inductance and capacitance. The gate driver loop area must be minimized to keep the inductance low and to reduce the emissions from the loop antenna created. The PCB trace connecting the source terminal of M1 with the GND pin of the HV9931 must be as short as possible to reduce inductance. The V_{DD} bypass capacitor, C_{DD}, provides the current for the MOSFET gate drive signal; it must have low ESR and located close to the V_{DD} pin of the HV9931.

Postconduction oscillation across D4 during the dead time of L1 can be another substantial source of RF emission. Adding a snubber circuit (R_d and C_d) can help significantly. In addition, this snubber is needed to reduce the voltage stress at D4 as discussed in Section 8.6.1.4.

The outside case of an electrolytic capacitor is connected to its negative terminal and in a traditional power supply this terminal is normally connected to the ground (0 V) rail to provide a degree of shielding. However, in the BBB circuit, this is not possible because the central energy storage capacitor C1 in this circuit is "floating" (not connected to circuit ground). It will behave like a radio frequency antenna. Placing the whole circuit into a shielded enclosure of some type is most likely necessary in the majority of applications.

8.7 LED Driver Design Example Using the BBB Circuit

Let us design a power converter for driving LEDs with the following characteristics:

Input AC line voltage: 80–260 V AC, 50–60 Hz
Output current: 350 mA
Output current ripple: ±15%
Output voltage: 35 V (max.)
THD: <20% at 120 V AC
Off-time: 15 μs

We will assume that the efficiencies of the input buck–boost stage and the output buck stage are $\eta_1 = 0.85$ and $\eta_2 = 0.9$ correspondingly. The efficiency of a DCM buck–boost stage is typically lower compared to the CCM buck stage. The overall efficiency $\eta = \eta_1 \times \eta_2 \approx 0.76$.

Note that a "universal" 80–260 VAC input range is not recommended, because of the high stresses on the components. Two designs, one covering 80–135 VAC and the other covering 150–260 VAC, would be the preferred option.

Step 1. Using the equation given earlier for T_{off}, we will calculate the timing resistor R_T value for $T_{off} = 10 \ \mu s$.

$$R_T = \frac{T_{off} - 880\,\text{ns}}{40 \cdot 10^{-12}} = \frac{1.412\,\mu s}{40 \cdot 10^{-12}} = 353\,\text{k}\Omega$$

Using the nearest standard resistor value, $R_T = 360\text{K}$.

Step 2. We will allow 30% peak-to-peak switching current ripple in $L2$, or $\Delta i_{L2} = 0.3 i_{L2} = 0.105$ A. Then the peak current in $L2$ is

$$i_{L2pk} = 0.4025 \ \text{A}.$$

The value of $L2$ can be calculated from

$$L2 = \frac{V_o \cdot T_{off}}{\Delta i_{L2} \cdot \eta_2} = \frac{35 \cdot 15 \cdot 10^{-6}}{0.105 \cdot 0.9} = 5.55\,\text{mH}$$

$L2 = 4.7$ mH is a standard value.

The DC current rating of $L2$ equals to $i_o = 0.35$ A. The saturation current rating of this inductor must be higher than the peak current.

$$i_{sat} > 0.4025 \ \text{A}.$$

Step 3. Assuming a 0.25-W power dissipation in the output current sense resistor R_{s2}, we can calculate its value.

$$R_{s2} = \frac{P}{i_o^2} = \frac{0.25\,\text{W}}{0.1225} = 2.0408\,\Omega$$

We will select a 1 Ω, 0.25 W resistor for R_{s2}, to minimize the power loss.

We can now calculate the value of R_{cs2}. Let us use the V_{DD} pin as a reference voltage ($V_{DD} = 7.5$ V). Although V_{DD} is reasonably accurate, it may drop during the AC line voltage cusps when there is no input voltage available at V_{in}, if the V_{DD} capacitor is too small to hold up the voltage. An external voltage reference is needed for better accuracy. Selecting $R_{ref2} = 100\text{K}$, we can calculate the value of R_{cs2} using the equation:

$i_{Lpk} = \dfrac{V_{ref} \cdot R_{cs}}{R_{ref} \cdot R_s}$, which transforms to

$$R_{cs} = \frac{i_{Lpk} \cdot R_{ref} \cdot R_s}{V_{ref}} = \frac{0.4025 \cdot 100 \cdot 10^3 \cdot 1}{7.5} = 5.366\,\text{k}\Omega$$

$R_{cs2} + R_{ov} = 5.366$ kΩ, let $R_{cs2} = 5.1$ kΩ and $R_{ov} = 270$ Ω.

Step 4. The input inductor $L1$ is assumed to reach BCM at $V_{AC\,min}$ at the peak of the input voltage hump. We can calculate the critical inductance value that meets this condition.

$$L1 = \frac{0.63 \cdot V_{AC\,min} \cdot \sqrt{2} \cdot T_{off}}{4 \cdot i_o} = \frac{56.7 \cdot \sqrt{2} \cdot 15 \cdot 10^{-6}}{1.4} = 859 \ \mu H$$

This is the maximum value, so we need to choose a lower inductance. Let $L1 = 700$ μH.

Step 5. Let us calculate the parameter δ and the duty cycle D at $V_{AC\,min}$, $V_{AC\,max}$, and V_{AC} using the equations:

Case 1, $V_{AC\,min} = 80$ V AC:

$$\delta_{min} = \frac{2 \cdot V_{AC\,min}^2 \cdot T_{off} \cdot \eta}{V_o \cdot i_o \cdot L1} = \frac{2 \cdot 6400 \cdot 15 \mu \cdot 0.85}{35 \cdot 0.35 \cdot 700 \mu} = 19$$

$$D = \frac{2 \cdot \left(\sqrt{1+\delta} - 1 \right)}{\delta} = 0.365$$

Case 2, $V_{AC\,max} = 260$ V AC:

$$\delta_{max} = \frac{2 \cdot V_{AC\,max}^2 \cdot T_{off} \cdot \eta}{V_o \cdot i_o \cdot L1} = \frac{2 \cdot 67,600 \cdot 15 \mu \cdot 0.85}{35 \cdot 0.35 \cdot 700 \mu} = 201$$

$$D = \frac{2 \cdot \left(\sqrt{1+\delta} - 1 \right)}{\delta} = 0.131$$

Case 3, V_{AC} at 120 V AC:

$$\delta = \frac{2 \cdot V_{AC}^2 \cdot T_{off} \cdot \eta}{V_o \cdot i_o \cdot L1} = \frac{2 \cdot 14,400 \cdot 15 \ \mu \cdot 0.85}{35 \cdot 0.35 \cdot 700 \ \mu} = 42.82$$

$$D = \frac{2 \cdot \left(\sqrt{1+\delta} - 1 \right)}{\delta} = 0.262$$

Step 6. The maximum peak current in $L1$ will occur at $V_{AC\,min}$. It can be calculated from:

$$i_{L1pk\,max} = \frac{V_{AC\,min} \cdot \sqrt{2} \cdot T_{off}}{L1} \cdot \frac{D}{1-D}$$

$$i_{L1pk} = 1.394 \text{ A}$$

Note that most "off-the-shelf" 700-µH DC chokes may be not suitable for $L1$. As the current in $L1$ cycles from zero to as high as i_{L1pk} every switching cycle, there may be excessive power dissipated in the magnetic core of $L1$ due to a large-magnetic flux excursion. On the other hand, the wire gauge used in such inductors is selected based on its DC current rating, whereas the RMS current in $L1$ is substantially lower than its peak current, so thin wire can be used. Thus, a custom designed $L1$ is likely to produce a more size-efficient solution. This makes prototyping more difficult, but in volume production there will be little impact in terms of cost.

Step 7. The next step is calculating the input current sense and divider resistors R_{s1} and R_{cs1}. Let us allow 0.1 W of power dissipation in R_{s1} at $V_{AC\,min}$. Power dissipation in R_{s1} can be calculated as:

$$P_{R_{s1}} = 0.1 = \frac{D_{max} \cdot i_{L1pk}^2 \cdot R_{s1}}{6}$$

Solving this equation for R_{s1}, we obtain:

$$R_{s1} = \frac{0.6}{D_{max} \cdot i_{L1pk}^2} = \frac{0.6}{0.365 \cdot 1.394} = 1.179\,\Omega$$

Let us select $R_{s1} = 0.47\ \Omega$, 1/4 W. To calculate R_{cs1}, we will use the equation

$$R_{cs} = \frac{i_{Lpk} \cdot R_{ref} \cdot R_s}{V_{ref}} = \frac{1.394 \cdot 100 \cdot 10^3 \cdot 0.47}{7.5} = 8.735\,k\Omega$$

This is assuming $V_{ref} = V_{DD}$ and $R_{ref1} = 100K$ as before. We will program the peak input current limit as 120% of i_{L1pk}, so that R_{cs1} is increased by 20%. Then,

$$R_{cs1} = 10.5\,k\Omega\ (10k\Omega\text{ would be suitable}).$$

Step 8. Let us assume the third harmonic distortion coefficient $K_3 = 0.15$ at $V_{AC} = 120$ V AC, 60 Hz. Then, the equation for the value of $C1$ is:

$$C1 = \frac{1}{\delta \cdot \left(1 + \dfrac{1}{\sqrt{1+\delta}}\right)} \cdot \frac{\eta_2 \cdot i_o}{\pi \cdot f_{AC} \cdot K_3 \cdot V_o}$$

$$C1 = \frac{1}{42.82 \cdot (1.151)} \cdot \frac{0.9 \cdot 0.35}{\pi \cdot 60 \cdot 0.15 \cdot 35} = 6.458\ \mu F$$

$$C1 \approx 10\ \mu F$$

Using the next equation at $V_{AC} = 260$ V AC, we can calculate the required voltage rating of $C1$.

$$V_C = \frac{V_o}{2 \cdot \eta_2} \cdot \left(1 + \sqrt{1+\delta}\right) = \frac{35}{2 \cdot 0.9} \cdot (1 + \sqrt{202}) = 295.8 \text{ V}$$

The voltage ripple at $C1$ is small at high-input voltage and can be ignored.

$$K_C = \frac{1}{\left(1 + \sqrt{1+\delta}\right)^2} \cdot \frac{\eta_2 \cdot i_o}{\pi \cdot f_{AC} \cdot C1 \cdot V_o}$$

$$K_{C\min} = \frac{1}{231} \cdot \frac{0.9 \cdot 0.35}{\pi \cdot 60 \cdot 10^{-5} \cdot 35} = 24.75 \text{ mV}$$

We would choose a voltage rating that gives some margin to allow for input voltage surges, so a 400-V rated capacitor is the next highest available rating.

The switching ripple current rating for this capacitor can be calculated at the minimum input voltage (80 V AC).

$$i_{CSW} = i_o \cdot \sqrt{\frac{64}{9 \cdot \pi \cdot \eta \cdot \eta_1} \cdot \frac{V_o}{V_{AC} \cdot \sqrt{2}} + D}$$

$$i_{CSW} = 0.35 \cdot \sqrt{\frac{64}{9 \cdot \pi \cdot 0.76 \cdot 0.85} \cdot \frac{35}{80 \cdot \sqrt{2}} + 0.365} = 0.42 \text{ A}$$

The RMS value of the second AC line harmonic is:

$$i_{C\text{line}} = \frac{\sqrt{2} \cdot i_o}{1 + \sqrt{1+\delta}} = \frac{0.4949}{5.472} = 0.09 \text{ A}$$

A 10 μF, 400 V capacitor rated for 0.51 A ripple current would be suitable.

Step 9. Optimal selection of the switching MOSFET M1 is based on finding a good balance between the total gate charge Q_g and the on-resistance RDS (ON). The drain voltage rating is given by the equation:

$$V_{DSM1} = V_{AC\max} \cdot \sqrt{2} + V_{C\max} \cdot \left(1 + K_C\right)$$

$$V_{DSM1} = 260 \cdot \sqrt{2} + 300 \cdot (1 + 0.025) = 675.2 \text{ V}$$

An 800-V MOSFET is required.

Acceptable gate charge Q_g is limited by the allowed power dissipation in the HV9931. The power dissipation can be estimated as:

$$P_{\text{reg}} = \left(\frac{2\sqrt{2} \cdot V_{AC\max}}{\pi} - V_Z \right) \cdot \left(\frac{Q_g \cdot (1 - D_{\min})}{T_{\text{off}}} + \frac{V_{\text{off}}}{V_{\text{off}1}} + \frac{V_{\text{off}}}{V_{\text{off}2}} + 1\,mA \right)$$

Where V_Z is Zener voltage of D10; let us use $V_Z = 47$ V. Let us select STD2N80 for M1. This is an 800 V, 1.8 A MOSFET by Fairchild with RDS (ON) = 6.3 Ω and $Q_{g\max} \approx 15$ nC at $V_{DS} = 640$ V, $V_{GS} = 10$ V.

Then:

$$P_{\text{reg}} = \left(\frac{2\sqrt{2} \cdot 260}{\pi} - 47 \right) \cdot \left(\frac{15 \cdot 10^{-9} \cdot (1 - 0.131)}{15 \cdot 10^{-6}} + \frac{7.5}{100 \cdot 10^3} + \frac{7.5}{100 \cdot 10^3} + 1\,mA \right)$$

$$P_{\text{reg}} = 377.5\,mW$$

The maximum power dissipation in HV9931LG (SO-8) is 900 mW at 25°C. However, above 25°C, the maximum power dissipation must be derated by 9 mW/°C. Thus, the maximum operating ambient temperature needs to be less than 83°C. A higher-voltage Zener diode can be selected to reduce power dissipation in the HV9931 and thus allow higher-ambient temperature. A higher-voltage Zener diode will also provide more protection against AC line voltage surges and will give some input undervoltage protection.

The maximum RMS current in M1 is calculated using:

$$i_{DM1} = \sqrt{\frac{D_{\max} \cdot i_{L1pk}^2}{6} + D_{\max} \cdot i_o^2} = \sqrt{\frac{0.365 \cdot 1.943}{6} + 0.365 \cdot 0.1225} = 0.404 \text{ A}$$

The peak current in M1 is

$$i_{L1pk} + i_{L2pk} = 1.394 \text{ A} + 0.4025 \text{ A} = 1.797 \text{ A}$$

Step 10. In accordance with the following equations, the average currents in D1–D4 are:

$$i_{D1} = \frac{4 \cdot \sqrt{2}}{\pi} \cdot \frac{i_o}{\eta_1 \cdot \left(1 + \sqrt{1 + \delta_{\min}}\right)} = \frac{1.98}{14.6125} = 0.136 \text{ A}$$

$$i_{D2} = D_{\max} \cdot i_o = 0.128 \text{ A}$$

$$i_{D3} = (1 - D_{\min}) \cdot i_o = 0.304 \text{ A}$$

$$i_{D4} = \frac{4 \cdot \sqrt{2}}{\pi} \cdot \left(\frac{2 \cdot \sqrt{2}}{\delta_{\min}} + \frac{1}{\eta_1 \cdot \left(1 + \sqrt{1 + \delta_{\min}}\right)} \right) \cdot i_o = 1.8 \cdot \left(\frac{2.8284}{19} + \frac{1}{4.6513} \right) = 0.655 \text{ A}$$

Peak currents in D1 and D4 equal to the peak current in $L1$, or:

$$i_{D1pk} = i_{D4pk} = i_{L1pk} = 1.394 \text{ A}.$$

The following equations give the reverse voltage across D1–D3, resulting in:

$$V_{RD1} = V_{ACmax} \cdot \sqrt{2} + V_C \cdot (1 + K_C) = 670.89 \text{ V}$$
$$V_{RD2} = V_{ACmax} \cdot \sqrt{2} = 367.7 \text{ V}$$
$$V_{RD3} = V_C \cdot (1 + K_C) = 303.2 \text{ V}$$

The minimum voltage rating for D4 if a good snubber is added in parallel will be:

$$V_{RD4} = V_{ACmax} \cdot \sqrt{2} = 367.7 \text{ V}$$

Thus in theory a fast 400-V diode can be used for D4 in a universal 80–260 V AC LED driver. In practice, a fast diode rated at 600 V is normally used, such as STTH1L06A (600 V, 1 A).

Adding an R_C snubber is recommended across D4. Reverse voltage across D4 depends on the capacitance value of C_D selected for this R_C snubber. The resistance and capacitance value of the snubber components depends mainly on the MOSFET output capacitance (the diode capacitance of D4 is much smaller and can be ignored). The typical data by Fairchild shows $C_{OSS} < 35$ pF at $V_{DS} > 50$ V for FQD2N80 (estimated $C_{OSS} \sim 20$ pF at V_{DS} greater than 100 V).

STTH108A by ST Microelectronics (800 V, 1 A, $t_{rr} = 75$ ns) can be selected for D1.

The snubber capacitor can be selected as (approximately)

$$C_D = 10 \cdot C_{OSS} = 220 \text{ pF}$$

And the snubber resistor is (approximately)

$$R_D = \frac{0.33}{C_D \cdot 2 \cdot \pi \cdot f_s} = 4.7 \text{ k}\Omega$$

Fast-switching rectifiers are needed for D2 and D3. We can select STTH1R06A (600 V, 1 A, $t_{rr} = 20$ ns) by ST Microelectronics.

Step 11. An output filter capacitor, C_o, of typically 100 nF will be needed for improved EMI performance. A larger-value capacitor can be used to reduce the ripple current in the LEDs further. This capacitor also acts as a bypass to any switching transients cause by parasitic circuit capacitance.

8.8 Buck With PFC

In 2016, Diodes Inc. introduced the AL1676. This is a high-voltage BCM buck converter and can operate from a rectified AC mains supply with a good PF. It has an internal MOSFET switch that is limited in current capability but sufficient for medium power LEDs, such as those with 100- or 150-mA current rating. The target application is retrofit lamps and tubes. A typical schematic is shown in Fig. 8.10.

This circuit operates in the BCM. This means that it is a peak current mode device that switches the MOSFET off when the current reaches twice the desired average current, and back on when the current has dropped to zero. Thus the average current is accurately controlled; this is shown in Fig. 8.11.

The peak current, i_{pk}, is set by resistor R_{cs}, which is connected between the current sense (CS) pin and ground (GND). Inside the AL1676, a comparator switches the MOSFET off when the voltage at the CS pin reaches 400 mV, which is the internal reference voltage. The resistor value is chosen to set the peak current level twice the desired average current by the equation:

$$R_{cs} = \frac{0.4}{2 \cdot i_{av}}$$

So for a 100-mA load, $R_{cs} = 0.4/0.2 = 2\ \Omega$.

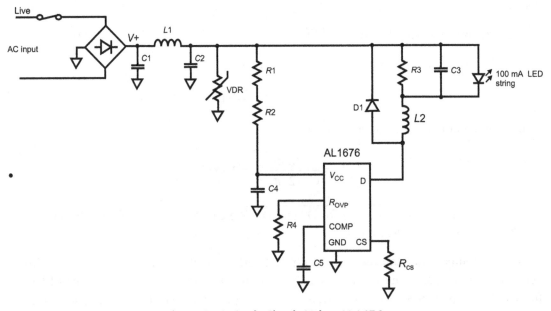

Figure 8.10: Buck Circuit Using AL1676.

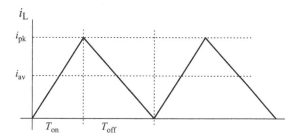

Figure 8.11: Inductor Current Waveform.

Choosing the correct inductor value is very important in this circuit. The AL1676 has a number of critical timing specifications:

$T_{\text{on min}} = 700$ ns, $T_{\text{on max}} = 29$ μs.
$T_{\text{off min}} = 6$ μs, $T_{\text{off max}} = 180$ μs.

The easiest to calculate is T_{off}: this is the time taken to discharge the inductor's energy, taking the current from i_{pk} to 0 mA. With a 100-mA load, $i_{\text{pk}} = 200$ mA. The inductor's load voltage is the forward voltage drop of the LEDs, which we can set at 60 V maximum in our example.

$$V = L\frac{\text{d}i}{\text{d}t}$$
$$\text{d}i = i_{\text{pk}} \text{ and } \text{d}t = T_{\text{off}}$$

To find the inductor value at $T_{\text{off min}}$

$$L = V \cdot \frac{T_{\text{off}}}{i_{pk}}$$

$L_{\text{min}} = 60$ V × 6 μs/0.2 A = 1.8 mH. Allowing for 20% tolerance, we get a value of 2.16 mH. The nearest standard value is 2.2 mH. So now we need to check T_{off}, using the revised inductor value of 2.2 mH.

$$T_{\text{off}} = L \cdot \frac{i_{\text{pk}}}{V_{\text{LED}}}$$

If we let $L = 2.2$ mH, $i_{\text{pk}} = 200$ mA and $V_{\text{LED}} = 60$ V (maximum), then $T_{\text{off}} = 7.33$ μs.

But what is the $T_{\text{on min}}$ period, when the AC mains voltage at its peak? If we take the peak AC mains voltage as $\sqrt{2} \times 265 = 375$ V, then the voltage across the inductor will be $375 - 60 = 315$ V.

$$T_{\text{on min}} = L \cdot \frac{0.2}{315} = 1.4\,\mu s$$

This is more than the minimum value of 700 ns, so 2.2 mH is a suitable inductor for a 60 V, 100 mA load.

Now consider an LED, such as the Osram Duris E5. This is specified at 120 mA, but would be suitable for our 100-mA load. Each LED has a maximum forward voltage drop of 2.8–3.6 V, so 16 LEDs in series would give us a load voltage of 57.6 V maximum, 51.2 V typical, or 44.8 V minimum. Consider the equations for T_{on} and T_{off} using the minimum voltage.

$$T_{off} = 2.2\,\text{mH} \cdot \frac{0.2}{44.8} = 9.82\,\mu s$$

$$T_{on\,min} = 2.2\,\text{mH} \cdot \frac{0.2}{330.2} = 1.33\,\mu s$$

These results verify that a 2.2-mH inductor is quite suitable.

The advantage of using BCM is that the PF is good, but the disadvantage is high-ripple current in the LED. Capacitor $C3$ across the LEDs helps to maintain LED current, even when the inductor current drops to zero.

Power Integrations is another manufacturer making a buck control IC that has a reasonable PFC performance, the LYT7503D. The control technique is described as critical conduction mode, which is basically the same as BCM used by the Diodes part AL1676 described earlier in this section. Reference designs for 120 V AC operation, using the LYT7503D, suggest that the circuits produce PFs of 0.9–0.95. These designs also allow triac dimming, although the design notes suggest that there may be flickering/shimmering present and give suggestions on how to reduce these effects.

8.9 Common Mistakes With PFC Circuits

The most common mistake is to use the wrong inductor for $L1$.

Inductors are sized both for their magnetic saturation level and for resistive heating. Thus an inductor may be specified as $i_{av} = 500$ mA, $i_{sat} = 400$ mA. This means that the inductor can pass 500 mA with a temperature rise of, say, 40°C. But it can only pass 400 mA before the inductance is reduced by, say, 10%. If this inductor were used in a PFC stage with a peak current of 400 mA it would overheat.

Inductor manufacturers do not normally specify magnetizing losses. The magnetic saturation levels are material dependent; the maximum flux density of ferrite is usually 200 mT, but can be higher for other materials. So when designing a ferrite-based inductor, a manufacturer will make the design based on this level. However, if the flux is changing from zero to a maximum level (such as in a PFC circuit), a significant amount of energy is lost in the ferrite core due to magnetization losses. The core loss can be considered a sort of magnetic resistance. So, a flux density limit of about 50–100 mT would be a better choice for a ferrite-based energy storage inductor in a PFC circuit, to prevent significant core heating due to these magnetization losses.

8.10 Conclusions

Detailed design analysis has not been given for the all of the PFC circuits. This chapter has been intended to show readers some options and point out limitations. For example, driving a single LED may require a BBB circuit because of the large step-down ratio, but longer strings can be driven from a Bi-Bred or a PFC stage followed by a buck converter.

Application notes from ST Microelectronics and Microchip cover the PFC, Bi-Bred, and BBB circuits in detail. These are proprietary and specialized solutions that are still evolving; interested readers should consult these application notes for the latest designs.

CHAPTER 9

Fly-Back Converters and Isolated PFC Circuits

Chapter Outline

A traditional fly-back converter uses an inductor with at least two windings (really, this is a transformer). Considering two windings; one is the primary, which is connected to the input power supply and a switch to ground; the other is the secondary, which is connected to the load. The circuit is arranged so that magnetic energy is stored in the inductor while switch is on; the current increases in the primary winding during this time. When the switch is off, the magnetic energy is released by current flowing out of the secondary winding. This is shown in Fig. 9.1.

The energy release is the "fly-back," so-called because in early television sets with a cathode ray tube, the fast rising voltage from a transformer winding was used to deflect the electron beam back to the starting point on the screen (the electron beam had to "fly back" quickly after completing a scan across the screen, to avoid missing the next line of data to be displayed).

Fly-back power supplies are relatively easy to design, but are more suited to constant voltage outputs. This is because the energy is stored in bursts, in a large reservoir capacitor at the output, and controlling the average voltage across the capacitor can be achieved with simple feedback. A constant current output is possible, but slightly more complex to design.

Fly-back converters are often used when the load must be isolated from the supply. Most fly-back circuits are used with AC mains input. By using constant on-time switching, they allow a good power factor (important in AC mains applications, as we saw in the last chapter).

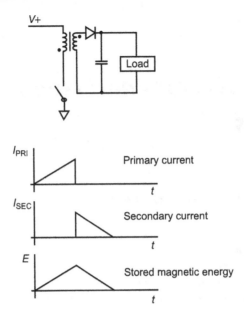

Figure 9.1: Fly-Back Principle.

Transformer-coupled circuits are less efficient than simple buck circuits and usually more costly, but if mechanical isolation from the AC mains is impossible, they are the only choice.

Some nonisolated fly-back circuits have a common ground connection, so that feedback can be directly applied to the control IC. Driving an isolated LED load is possible if the secondary winding is completely isolated from the primary winding. Some general-purpose applications can use simple current limit techniques in the primary winding to control the output current from the secondary winding: basically controlling the power only. More accurate control will either require an optocoupler for the feedback signal, or primary side control. In primary side control, signals reflected back from the load cause modulation on the primary winding (when the primary switch is not conducting) and these signals are used for feedback.

Two semiconductor companies dominate the supply of control ICs for both isolated and nonisolated fly-back converters. One company is iWatt, now part of Dialog Semiconductor, with parts like the IW3688. The other major supplier is Power Integrations, with parts like the LYT4322E. Both of these driver ICs enable good PFC and triac phase dimmer operation, which are ideal characteristics for light bulb replacements.

9.1 Single-Winding Fly-Back (Buck–Boost)

Some fly-back converters use an inductor with a single winding. These are buck–boost controllers and are an alternative to the boost–buck converters like Ćuk and SEPIC types that were discussed in Chapter 7. Clearly, isolation is not possible with this type of converter

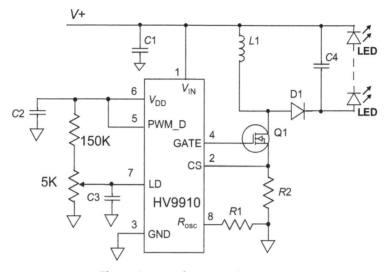

Figure 9.2: Buck–Boost Converter.

because a single inductor winding is used for both the primary and secondary currents. This is shown in Fig. 9.2.

Current is forced through the inductor when the MOSFET switches on and connects the inductor across the power supply rail. The current level rises almost linearly with time. At a predetermined current level, the MOSFET is turned off and, since the current flow cannot stop instantly, it is forced to flow through a diode to charge the output capacitor and drive the load. The current in the inductor falls back to zero as the energy stored in the magnetic core is depleted. The single-winding fly-back can be calculated from the number of volt-seconds on the charge cycle equaling the number of volt-seconds on the discharge cycle.

The duty cycle of a buck–boost converter (continuous conduction mode) is given by the equation:

$$\frac{V_O}{V_I} = \frac{D}{1-D}$$

$$V_O \cdot (1-D) = V_I \cdot D$$

$$V_O = V_I \cdot D + V_O \cdot D = D \cdot (V_I + V_O)$$

$$D = \frac{V_O}{V_I + V_O}$$

So, if we have $V_{IN} = 24$ V and $V_{OUT} = 30$ V, $D = 30/54 = 0.555$.

In practice we want discontinuous conduction mode because continuous conduction mode is difficult to stabilize. This means that the inductor current falls to zero at the end of each cycle. So, assume we want 350 mA output and 100 kHz switching frequency. The period is 10 μs, so the on-time is 5.55 μs and the off-time is 4.45 μs. During the off-time, the current in the inductor falls linearly from a peak level to zero. To average 350 mA output, the average inductor current during the off-time must be 350/0.445 mA = 786.5 mA, so the peak current must be double this, or 1.573 A. This also means that during the on-time, the current must rise from zero to 1.573 A.

The voltage from the power supply is 24 V, so using the familiar equation:

$$E = -L \cdot \frac{di}{dt}$$

$$L = E \cdot \frac{dt}{di} = 24 \cdot \frac{5.55 \cdot 10^{-6}}{1.573} = 84.67 \, \mu H$$

In practice there should be some dead time allowed, when the inductor carries no current, to ensure discontinuous conduction mode. This dead time is to allow for power supply tolerances, inductor tolerances, etc. Too much dead time means that the peak current is higher and this reduces the efficiency of the power supply.

Suppose we allow 25% tolerance, so that the on-time is 4.44 μs; this will reduce the inductance by 25%.

$$L = E \cdot \frac{dt}{di} = 24 \cdot \frac{4.44 \cdot 10^{-6}}{1.573} = 68 \, \mu H$$

The off-time will be reduced unless the peak current is increased in proportion.

$$E = -L \cdot \frac{di}{dt}$$

$$-30 = 68 \cdot 10^{-6} \cdot \frac{di}{4.45 \cdot 10^{-6}}$$

$$di = \frac{-30 \cdot 4.45 \cdot 10^{-6}}{68 \cdot 10^{-6}} = 1.963 \, A$$

Increasing the peak current by 25% gives the desired result. The peak current is set by the value of current sense resistor between the MOSFET source and ground.

9.2 Two-Winding Fly-Back

A schematic of a typical fly-back circuit for driving LEDs is shown in Fig. 9.3. The dot alongside the transformer winding indicates the start of the winding. In this case the start is connected to the MOSFET drain, which alternates between a ground connection and open circuit. The voltage at the drain, and hence the winding start point, varies considerably during switching. Conversely, the outer layer (end of the winding) is at a fixed high voltage. The high voltage rail is highly coupled to ground with multiple capacitors, so that for AC signals it is effectively a ground node. Having the outer winding connected to a ground node helps to screen the inner layers, which reduces radiated EMI.

The secondary-winding start point is connected to the output diode, which prevents conduction when the MOSFET is on. The start point of the secondary winding is connected to the anode of the output diode. Energy that is stored during the MOSFET on-time is released during the off-time, by current flowing through the output diode and into the load.

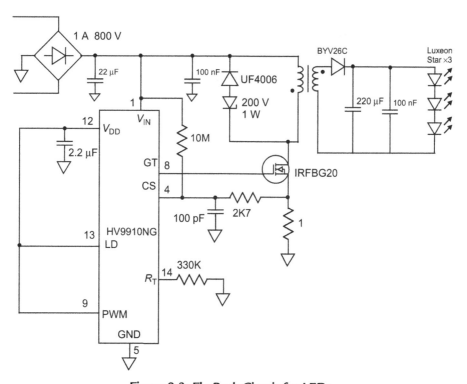

Figure 9.3: Fly-Back Circuit for LEDs.

Calculation of the transformer characteristics, like inductance value and primary to secondary turns ratio, are very important in the design. In order for complete power transfer from the primary to secondary, the volt-seconds must be equal. The equation is:

$$\frac{V_{PRI} \cdot T_{ON}}{N_{PRI}} = \frac{V_{SEC} \cdot T_{OFF}}{N_{SEC}}$$

9.2.1 Fly-Back Example

Let us make an isolated 3-W lamp by connecting three white power LEDs in series.

Suppose we have a primary voltage of 48 V and an on-time of 5 μs, and the primary to secondary turns ratio is 1:0.1. If we are driving a 10-V LED load, the off-time will be given by transposing the equation

$$\frac{V_{PRI} \cdot T_{ON}}{N_{PRI}} = \frac{V_{SEC} \cdot T_{OFF}}{N_{SEC}}, T_{OFF} = \frac{V_{PRI} \cdot T_{ON} \cdot N_{SEC}}{N_{PRI} \cdot V_{SEC}} = \frac{48 \cdot 5 \cdot 10^{-6} \cdot 0.1}{1 \cdot 10} = \frac{24 \cdot 10^{-6}}{10}$$

Thus T_{OFF} = 2.4 μs. Thus the switching period must be greater than 7.4 μs to allow complete removal of the magnetic energy in the transformer core. A switching frequency of below 130 kHz will be satisfactory, let us use 100 kHz to give some margin.

With 100 kHz switching, the period will be 10 μs. If the average output current is 350 mA, the average in 2.4 μs will be 1.46 A (350 mA * 10/2.4 = 1.46 A). Since this current decays linearly from the transformer winding, the peak secondary current will be double this: 2.92 A. The secondary inductance will be

$$E = -L \cdot \frac{di}{dt}.$$

$$L = E \cdot \frac{dt}{di} = 10 \cdot \frac{2.4 \cdot 10^{-6}}{2.92} = 8.22 \, \mu H$$

Since the primary has 10 times the turns of the secondary, the primary inductance will be 100 times that of the secondary (the turns ratio, N, is squared when calculating inductance). In other words, the primary inductance will be 822 μH.

Most current-mode power supplies control the switching so that the MOSFET turns off when a certain peak current is reached in the primary winding. Since the peak current in the secondary is 2.92 A and the turns ratio is 10:1, we need a peak primary current of 292 mA. [Check: $E = -L \cdot \frac{di}{dt}$, so E = 822 * 10^{-6} * 0.292/(5 * 10^{-6}) = 48 V].

The problem with the design that we have is that the LED current will change if the LED voltage changes because we have based our design on a certain output voltage. Actually this gives a constant power output, assuming a constant voltage input, which is fine for noncritical designs. But what if the input voltage changes?

A higher input voltage will mean that the current limit will be reached in a shorter time. This means that the duty cycle will be reduced and hence the number of volt-seconds on the primary will be unchanged. In practice, inherent delays in the current sense comparator will cause the input current to "overshoot" the reference level. This overshoot is greater with increasing input voltage because the delay is constant but the rate of current rise is increasing with input voltage. Compensation of this overshoot can be achieved by connecting resistor between the supply voltage rail and the current sense pin. This resistor injects a small DC bias that increases with increasing supply voltage and thus triggers the comparator earlier as the supply voltage rises.

The 10:1 turns ratio and 10 V output used in the previously mentioned example cause a reflected voltage of 100 V in the primary winding when the secondary conduction takes place. This reflected voltage adds to the supply voltage, so a MOSFET with a 200 V or higher voltage rating is required when powering this circuit from a 48-V supply.

The design example does not allow for efficiency. In practice a fly-back converter has about 90% efficiency, so the input current must be increased by about 10% to allow this.

If we were designing a constant voltage circuit, we would allow the peak primary current to be higher than that given in the example. This margin allows for the input voltage variations. We would then use feedback from the output to control the switching, to reduce the power in the primary, as necessary.

9.3 Three-Winding Fly-Back

Some fly-back power supplies use a third winding, called a bootstrap or auxiliary winding, as shown in Fig. 9.4. This is used to power the control IC, once the circuit is operating. The bootstrap winding has the same orientation as the secondary winding and the voltage is simply determined by the turns ratio of the bootstrap compared to the secondary. In our example of a 10-V output from the secondary, the bootstrap could have the same number of turns and thus give (approximately) 10 V for powering the control IC.

At start-up, there is no power available from the bootstrap winding, so a start-up regulator is required. In the circuit shown in Fig. 9.4, a Microchip HV9120 is shown, which has an internal start-up regulator. However, many PWM controllers operate from low voltage and thus need an external start-up circuit. Example start-up regulators are the LR645 and the LR8 from Microchip; these give a low voltage and low current output from an input with a voltage as high as 450 V. Once the bootstrap produces power, the start-up regulator turns off. This is more efficient than using a dropper resistor and Zener diode as a start-up circuit.

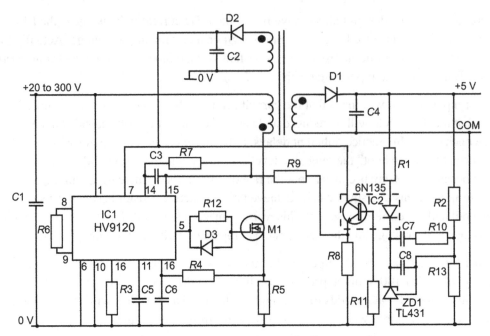

Figure 9.4: Fly-Back Using A Three-Winding Transformer.

9.3.1 Design Rules for a Fly-Back Converter

This section gives design rules for a fly-back converter based on either turns ratio selection determined by the maximum duty cycle allowed (Case 1), or by the optimum turns ratio based on the maximum working voltage of the MOSFET switch (Case 2). In Case 1, a design based on the maximum duty cycle (at the lowest input voltage) allows the widest input voltage range. In Case 2, a design based on the maximum voltage across the MOSFET allows a potentially lower cost solution. Alternatively, a fly-back design based on an already available transformer with a known (and fixed) turns ratio may be considered.

The transfer function of a fly-back converter is:

$$\frac{V_O}{V_I} = \frac{D}{(1-D)} \cdot N$$

So the duty cycle can be found by transposing the equation:

$$V_O \cdot (1-D) = V_I \cdot D \cdot N$$

$$V_O = V_I \cdot D \cdot N + V_O \cdot D = D \cdot (V_I \cdot N + V_O)$$

$$D = \frac{V_O}{(V_I \cdot N) + V_O}$$

9.3.2 Case 1: Turns Ratio Based on Maximum Duty Cycle

Given the minimum input voltage $V_{\text{I\_MIN}}$, output voltage V_O, and maximum duty cycle D_{MAX}, the turns ratio N can be calculated:

$$N = \frac{V_O \cdot (1 - D_{\text{MAX}})}{V_{\text{I\_MIN}} \cdot D_{\text{MAX}}}$$

D_{MAX} is typically chosen as 45% (0.45) for a PWM controller with a maximum 49% duty cycle.

If we take the earlier example of 48-V input (say, 46 V minimum), 10-V output (add 0.6 V for the output diode), and allow 45% duty cycle, we get:

$$N = \frac{10.6 \cdot (0.55)}{46 \cdot 0.45} = 0.282$$

This is the minimum value. A transformer with a convenient turns ratio of 1:0.33 (3:1) could be used. The maximum duty cycle will then be:

$$D = \frac{V_O}{(V_I \cdot N) + V_O} \frac{10.6}{(15.33 + 10.6)} = 0.41 \ \ (41\%)$$

9.3.3 Case 2: Turns Ratio Based on Maximum Switch Voltage

The output voltage across the secondary winding is induced into the primary and magnified by the turns ratio N. This was illustrated at the beginning of this chapter, when a 10 V output caused 100 V to be induced into the primary winding of a 10:1 turns ratio transformer. Considering that the supply voltage was only 48 V, this forced us to use a 200-V MOSFET as the primary switch. The aim here is to minimize the MOSFET switch operating voltage requirement.

As the voltage reflected into the primary often has some ringing, a snubber circuit is used to limit the voltage across the primary winding. Ringing is due to resonance between the MOSFET drain capacitance, parasitic capacitance in the circuit, and parasitic inductance of the transformer primary. Parasitic inductance in the transformer is often referred to as "leakage inductance" because it is the proportion of the primary inductance that is not coupled into the secondary, so the magnetic field "leaks out."

A Zener diode is sometimes used as a snubber. The voltage across the Zener diode will be greater than the voltage induced into the primary from the secondary (output) voltage, otherwise power dissipation and losses will both be very high.

$$V_O = N \cdot (V_{\text{SW}} - V_Z - V_{\text{IN\_MAX}})$$

To find the secondary-winding voltage, the forward voltage drop of the output diode, V_F, must be added to the output voltage.

$$N = \frac{V_O + V_F}{(V_{SW} - V_Z - V_{IN\_MAX})}$$

As a safety margin, $(V_{SW} - V_Z - V_{IN\_MAX}) \geq 10\,\mathrm{V}$.

In the example we used earlier, with 48 V input, we could have used a 100-V switch and a 33-V Zener diode. The output is 10 V, so allowing for V_F this becomes 10.6 V across the secondary winding:

$$N = \frac{10 + 0.6}{(100 - 33 - 48)} = \frac{10.6}{19} = 0.558$$

We could use a transformer with 2:1 turns ratio ($N = 0.5$). The primary voltage induced from the secondary winding will be 21.2 V, which is below the Zener diode voltage by 11.8 V, which is a reasonable margin to minimize power dissipation. The peak voltage across the MOSFET drain will be limited to 48 V + 33 V = 81 V.

With a turns ratio of 2:1, the maximum duty cycle with a 46 V minimum input voltage will be:

$$D = \frac{V_O}{(V_I \cdot N) + V_O} = \frac{10.6}{23 + 10.6} = 0.315 \quad (31.5\%)$$

9.3.4 Inductance Calculations

Now as we have the turns ratio (by either means described earlier) and the maximum duty cycle, we can determine the inductance and switching current. Let us use Case 1, with 41% as the maximum duty cycle.

$$P_{IN} = \frac{P_{OUT}}{\eta}$$

The output power is 10 V $\times$ 0.35 A = 3.5 W and the efficiency can be guessed at as being 85%. The input power is then 4.12 W. The input current at minimum input voltage is then:

$$I_{AV} = \frac{P_{IN}}{V_{IN}} = \frac{4.12}{46} = 0.09\,\mathrm{A}$$

$$I_{PK} = \frac{2 \cdot I_{AV}}{D_{MAX}}$$

At 46 V_{IN} and 41% duty cycle:

$$I_{PK} = \frac{2 \cdot 0.09}{0.41} = 0.439 \text{ A}$$

With 60 kHz switching, the period will be 16.667 µs. With a 41% duty cycle, the switch on-time will be 6.835 µs. So we need the primary current to rise by 439 mA in 6.835 µs.

$$L_{PRI} = \frac{V_{IN} \cdot dt}{di} = \frac{46 \cdot 6.835 \cdot 10^{-6}}{0.439} = 716 \text{ µH}$$

The secondary has 1/3 the number of turns compared to the primary, so the inductance of the secondary will be 1/9, or 79.55 µH.

The other design parameter for the transformer is the size and A_L factor of the ferrite core. The A_L factor is defined as the number of nano-Henries inductance per turn (nH), of the winding wire. Inductance is then given by $L = A_L \cdot N^2$, where N is the number of turns. In a fly-back transformer an air gap between the two halves of the ferrite core is necessary to prevent magnetic saturation; as the air gap increases, the A_L factor reduces. The flux density (B) will depend on the cross-sectional area of the core (A_C), given in square meters. Suppose in this case we have some E20 cores available from Ferroxcube. For E20/10/6 cores, the core cross-sectional area is 32 mm². So $A_C = 32 * 10^{-6}$ m². The number of turns can be calculated, based on the design parameters mentioned earlier and using $B = 200$ mT as the maximum flux density:

$$N = \frac{L_{PRI} \cdot I_{PK}}{A_C \cdot B_{MAX}} \text{ (turns)}$$

$$N1 = \frac{716 \cdot 10^{-6} \cdot 0.439}{32 \cdot 10^{-6} \cdot 0.2} = 49$$

$$A_L = \frac{L_{PRI}}{N1^2} = \frac{716 \cdot 10^{-6}}{2401} = 298 \text{ nH}$$

Refer to core manufacturers' specifications and chose a core with a lower A_L value (larger gap) than calculated using the previous equation. A suitable core (3C90 material, 160 µm air gap) has an A_L value, 250 nH. The number of turns can then be calculated as:

$$N = \sqrt{\frac{L}{A_L}}, \text{ where } L \text{ is expressed in nH. Thus } 716 \text{ µH} = 716,000 \text{ nH.}$$

$N_{PRI} = 54$ (rounding up to the next highest value. This quite conveniently gives the secondary turns as $N_{SEC} = 18$, since it is 1/3.

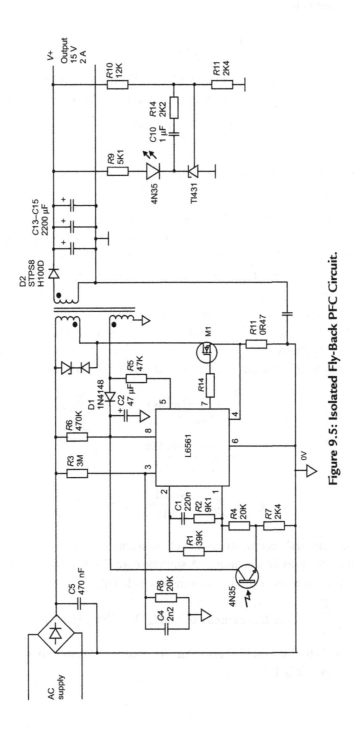

Figure 9.5: Isolated Fly-Back PFC Circuit.

9.4 Three-Winding Fly-Back PFC

An isolated fly-back is shown in Fig. 9.5 (based on a design in ST Microelectronics application note AN1060). This uses a coupled inductor, which has three windings on the same magnetic core: primary, secondary, and auxiliary. For the circuit description, we can ignore the auxiliary winding used to power the controller IC. The primary winding is on the input side and the secondary winding is on the output side. When the MOSFET turns on, current increases in the primary winding, storing magnetic energy in the magnetic core. Constant on-time switching ensures that the stored energy rises and falls as the AC mains voltage rises and falls, to give a good power factor.

When the MOSFET turns off, after a fixed time, the magnetic energy is released by current flowing in the secondary winding. The secondary current charges three large capacitors, which provides a constant DC voltage for the load.

The secondary voltage can be higher or lower than the supply voltage, as the AC supply voltage rises and falls, hence the use of the term boost–buck. In this case, the output is a constant voltage of 15 V and up to 2 A current, but these levels could be changed to suit the application. For LED driving, a number of constant current buck circuits could be powered from the 15 V output.

Essentials of Switching Power Supplies

This chapter will examine the advantages and disadvantages of the various driver techniques, which have already been described. The issues of efficiency, electromagnetic interference (EMI), cost, and other requirements that are additional to the basic function of the LED driver.

10.1 Linear Regulators

In Chapter 4 we saw how the use of linear regulators can cause heat dissipation problems because of low efficiency. On the other hand, "switched linear" regulators were described for AC mains applications, which had a very respectable efficiency combined with a good power factor.

A linear LED driver is generally less efficient than a switching driver, but sometimes a linear driver can be more efficient. For example, if a 12-V power source and three LEDs, each having a 3.5-V forward drop are present, by connecting them in series the total drop is 10.5 V. The efficiency of a linear LED driver, dropping only 1.5 V will be 87.5%. This would be a respectable efficiency for a switching LED driver. And for a linear regulator there is no EMI to be filtered.

On the other hand, driving one LED from a 12-V supply would give an efficiency of 3.5/12 = 29% with a linear LED driver. Here a buck switcher would give closer to 90% efficiency (Fig. 10.1). Efficiency is important where heat dissipation must be minimized. Otherwise cost usually takes precedent and the cost of a switching regulator with EMI filters would be somewhat higher.

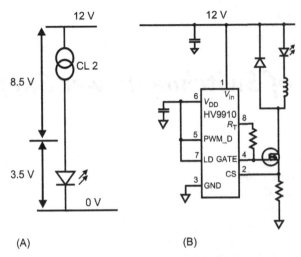

(A)   (B)

Figure 10.1: Linear Versus Switching Solutions.
(A) <30% Efficient. (B) >90% Efficient.

10.2 Switching Regulators

In Chapters 5–9 we looked at switching regulators, which generally have much higher efficiency compared to linear regulators, but can generate EMI that has to be suppressed by careful circuit board design, screening, and filtering. This is a legal requirement and product cannot be sold unless the equipment meets the standards laid down in law. The EMI-reducing techniques are described in Chapter 13.

Conversely, where EMI requirements are very demanding, such as medical and automotive applications, linear LED driver techniques can be used instead. Of course the efficiency may suffer, and so a heat sink will be needed, but this is sometimes much better than trying to make a switching circuit in terms of cost and physical size.

Although Microchip's LED driver–integrated circuits are used in examples throughout this book, most examples can be adapted to use similar drivers from other manufacturers. For example, the Linear Technology LTC3783 has similar functions to the Microchip HV9911. The National Semiconductor LM5020 is a buck controller, like the HV9910. However, Microchip devices have an internal high-voltage regulator, which makes them more versatile. Having said that, several manufacturers have made LED drivers with high-voltage regulators, based on the HV9910 (some remarkably similar!).

Why should we choose a buck converter scheme? Or a buck–boost? Or a boost? Or any other scheme?

- The simplest scheme is the buck and this should be used if the LED string voltage is no more than about 80% of the supply voltage (85% absolute maximum). This is to allow sufficient voltage across the inductor, current sense resistor, and MOSFET switch. The minimum off-time is needed for operation of the regulation circuits.

• A boost scheme should be used if the LED string voltage is always 150% or more of the supply voltage (120% absolute minimum). This is to limit the duty cycle to about 85%, allowing sufficient MOSFET switch off-time.

• A boost–buck should be used if the LED string voltage is close to the supply voltage, or can overlap it. This topology is often used if the supply voltage varies a lot over time, such as in battery-powered equipment. If an AC input with power factor correction (PFC) or isolation is required, boost–buck schemes using SEPIC or fly-back topologies should be considered.

10.2.1 Buck Regulator Considerations

In Chapter 5 we first looked at the simplest switching regulator, the buck converter. In a buck circuit the load voltage must be less than 85% of the supply voltage, otherwise the output becomes difficult to control. Buck circuits are used for mains-powered LED drivers, when driving a long string of LEDs. Buck circuits are also used where the input supply voltage is relatively low, say in a 12 V DC–automotive application, but where just one LED is being driven.

Buck regulators can be very efficient, maybe 90–95%, especially if the load is a long string of LEDs with a moderately high-forward voltage (i.e., high-duty cycle). This is because the power dissipation in the flywheel diode is a smaller proportion of the total power because the flywheel diode only conducts during the MOSFET off-time, which is a smaller proportion of the total switching cycle. The MOSFET dissipates power during the on-time, when it is conducting, but the voltage drop across the MOSFET switch is usually much lower than the forward drop of a fast rectifier. Synchronous buck regulators, which use a MOSFET in place of the flywheel diode, are sometimes used in low voltage applications and their efficiency can be higher than 95%.

To operate correctly there must be some ripple in the output current. The output current needs to reduce enough to allow the current sense comparators to be reset. The output ripple current Δi_o is normally designed to be 20–30% of the nominal output current, i_o. This ensures that the output current falls far enough in each cycle so that noise in the current sense comparator has little effect. If the ripple current is below 10% of i_o, the switching of the MOSFET can be erratic.

The peak current in the LED string is $i_{o\,max} = \dfrac{V_{th}}{R_{sense}}$

The average output current in the LED string (i_o) is given by the equation:

$$i_o = \frac{V_{th}}{R_{sense}} - \frac{1}{2} \cdot \Delta i_o$$

Here V_{th} is the current sense comparator threshold, and R_{sense} is the current sense resistor. The average current is the peak current minus half the ripple.

The ripple current can introduce a peak-to-average error in the output current setting that needs to be accounted for. When the constant off-time control technique is used, the ripple current is nearly independent of the input supply voltage variation. Therefore, the output current will remain unaffected by the varying input voltage.

Adding a filter capacitor across the LED string can reduce the output current ripple in the LED, without reducing the inductor ripple current. This means that a lower-value inductor could be used without increasing the LED ripple current. However, keep in mind that the peak-to-average current error is affected by the variation of the MOSFET off-time, T_{off}. Therefore, the initial output current accuracy might be sacrificed with large-ripple current levels in the inductor.

Alternatively, if the inductor value is unchanged, adding a filter capacitor across the LED terminals allows an apparently more "constant" current in the LED. This capacitor also reduces EMI at the output by providing a bypass path for any switching current spikes. Switching current spikes could also reduce the LED lifetime, so a filter capacitor reduces this risk.

Another important aspect of designing an LED driver is related to certain parasitic elements of the circuit. These include distributed coil capacitance of the inductor C_L, junction capacitance, C_J, and reverse recovery current in the flywheel diode, as well as capacitance of the printed circuit board traces C_{PCB} and output capacitance C_{drain} of the MOSFET. These parasitic elements affect the efficiency of the switching converter because they cause switching losses. These parasitic elements are shown in Fig. 10.2.

Parasitic elements cause a switch-on current spike that could potentially cause false triggering of the LED driver IC's current sense comparator. This can be mitigated against by an RC filter fitted between the MOSFET source and the current sense (CS) pin. Minimizing parasitic elements is essential for efficient and reliable operation of the buck converter.

Coil capacitance of inductors is typically provided in the manufacturer's data books either directly or in terms of the self-resonant frequency (SRF).

$$\text{SRF} = 1/(2\pi\sqrt{L \cdot C_L}),$$

Here L is the inductance value, and C_L is the coil capacitance. Charging and discharging this capacitance every switching cycle causes high-current spikes in the LED string. Therefore, connecting a small capacitor C_o ($\sim$10nF) across the LED string is recommended to bypass these spikes, as mentioned earlier.

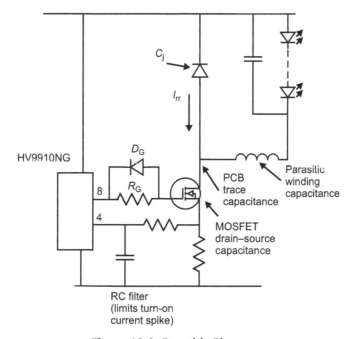

Figure 10.2: Parasitic Elements.
C_J, Junction capacitance; I_{rr}, reverse recovery current; *PCB*, printed circuit board.

Using an ultrafast rectifier flywheel diode is recommended to achieve high efficiency and reduce the risk of false triggering of the current sense comparator. When the MOSFET turns on, the flywheel diode's state changes from "forward conduction" to "off" (reverse bias). This change of state in the flywheel diode cannot happen immediately because electrical charges have to move from one place to another inside the semiconductor material, which takes time. There is always a reverse recovery current flowing in the opposite direction for a short period, t_{rr}, during this change of state.

Using diodes with shorter reverse-recovery time, t_{rr}, and lower junction capacitance, C_J, improves performance. In low-voltage applications a Schottky diode can be used, but in high-voltage applications a diode with less than 75-ns reverse recovery time (preferably less than 50 ns) should be used. There are high-voltage Schottky diodes, called silicon carbide (SiC), but these are high in cost.

The reverse voltage rating V_R of the diode must be greater than the maximum input voltage of the LED lamp. The forward voltage drop of diodes with very fast-recovery times is sometimes relatively high and can lead to high-conduction losses, so also consider this when making a diode selection. Low-voltage Schottky diodes have a typical forward voltage between 0.4 and 0.5 V, but this varies between types and the designer should check the specific parts' datasheet.

The total parasitic capacitance present at the DRAIN output of the MOSFET can be calculated as:

$$C_P = C_{drain} + C_{PCB} + C_L + C_J$$

When the switch turns on, the total parasitic capacitance C_P is discharged into the DRAIN output of the MOSFET. The discharge current amplitude is limited to the MOSFET saturation current level, so MOSFETs with a high–on-resistance and a lower-saturation current can sometimes produce lower overall losses. This is especially true if the duty cycle is small because the switch is conducting for a small proportion of the time and hence the conduction losses will not be significant. Note that the saturation current in a MOSFET becomes lower at increased junction temperature.

The duration of the leading edge current spike can be estimated as:

$$t_{spike} = \frac{V_{in} \cdot C_P}{i_{sat}} + t_{rr}$$

To avoid false triggering of the current sense comparator, C_P must be minimized in accordance with the following expression:

$$C_P < \frac{i_{sat} \cdot (t_{blank\,min} - t_{rr})}{V_{in\,max}}$$

The factor $t_{blank\,min}$ is the minimum blanking time, which depends on the control IC and is in the order of 300 ns. When the MOSFET gate drive is activated, the control IC disables the current sense input for this time period, to avoid false triggering from the switch-on current surge, previously described. The factor $V_{in\,max}$ is the maximum instantaneous input voltage.

Discharging the parasitic capacitance C_P into the DRAIN output of the MOSFET is responsible for the bulk of the switching power loss. It can be estimated using the following equation:

$$P_{switch} = \left(\frac{C_P V_{in}^2}{2} + V_{in} i_{sat} \cdot t_{rr} \right) \cdot f_s$$

where f_s is the switching frequency, i_{sat} is the saturated DRAIN current of the MOSFET. The switching loss is the greatest at the maximum input voltage.

The switching frequency of a buck converter having constant off-time operation is given by the following:

$$f_s = \frac{V_{in} - \eta^{-1} \cdot V_o}{V_{in} \cdot T_{off}}$$

where η is the efficiency of the power converter. This value for f_s based on typical values for V_{in} and V_o can be used in the previous equation if a value of constant switching frequency is not available.

The switching power loss associated with turn-off transitions of the DRAIN output can be disregarded. Due to the large amount of parasitic capacitance connected to this switching node, the turn-off transition occurs essentially at zero voltage.

Conduction power loss in the MOSFET can be calculated as

$$P_{cond} = D \cdot i_o^2 \cdot R_{on}$$

where $D = V_o/(\eta \cdot V_{in})$ is the duty ratio, R_{on} is the ON resistance.

10.2.1.1 Buck Converter AC Input Stage

An off-line LED driver requires a bridge rectifier and input filter; the design of the input filter is critical for obtaining good EMI.

We may use an aluminum electrolytic capacitor after the bridge rectifier, to prevent interruptions of the LED current at zero crossings of the input voltage (the cusps in the rectified sine wave, or "haversine," waveform). As a "rule of thumb," 2~3 µF per watt of the input power is required. An electrolytic capacitor is often used and has the added ability of absorbing voltage surges that may be present on the AC line. However, consider the power factor requirements of the system. A simple rectifier–capacitor solution may be unacceptable for many applications.

Large values of input capacitor will cause unacceptably high-current surges when power is first applied. These current surges can damage the electrolytic capacitor, reducing its life expectancy, and also damage the switch or electrical connectors at the AC line. Inrush current limiters are often connected in series with the AC line to prevent the current surge; see Section 10.2.6.

An inductor in series with the supply rail, after the input capacitor, is needed to present high impedance to switching frequency signals, as shown in Fig. 10.3. The current rating of this inductor needs to be higher than the expected current level in normal operation, to allow for the switch-on surges. The value of the inductor depends on the level of signal attenuation required, when combined with the input capacitor shunt impedance, to meet the required EMI standards.

The impedance of an inductor is given by: $X_L = 2 \cdot \pi \cdot f_s \, L$, so if we needed 200-$\Omega$ impedance at 100 kHz to give us our desired attenuation, $L = 0.318$ mH, a 330-µH filter inductor could be used.

A capacitor connected between the switching side of the filter inductor and ground, albeit of small value, is necessary to ensure low impedance to the high-frequency switching currents

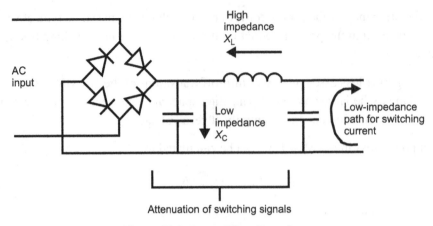

Figure 10.3: Input Filter Functions.

of the converter. As a rule of thumb, this capacitor should be approximately 0.1–0.2 μF/W of LED output power. A 100-nF capacitor can be used in a circuit that drives a single 1-W LED.

10.2.2 Boost Regulator Considerations

The output voltage in a boost circuit must always be higher than the input voltage by about 20% or more, and this was discussed in Chapter 6. Ignoring PFC applications, a boost converter–driving LED will always be powered from a low-voltage DC supply. For example, the backlight in a cell phone with a color LCD display usually employs low-cost white light LEDs. A boost regulator is used in this application to drive a string of 20-mA LEDs from a 3–4 V battery.

As another example, in flat-screen television backlighting high-power red, blue, and green (RGB) LEDs are used to create a white light that exactly matches the LCD and produces true colors. In this application a boost converter powered from a 12- or 24-V DC supply is used to drive many 350-mA LEDs connected in series, with a forward voltage in the range 40–80 V.

Boost regulators should always be provided with overvoltage protection, in case the LED load is disconnected. Otherwise the output voltage will continue to rise and eventually cause component breakdown. In safety electrical low-voltage systems, the output voltage would normally be kept below 42 V.

10.2.3 Boost–Buck Regulator Considerations

To operate in an environment where the input voltage could be higher or lower than the output voltage, a buck–boost (or boost–buck) circuit is necessary. Boost–buck circuits were described in Chapter 7. The situation of having a load voltage range that overlaps the supply voltage range is commonly found in automotive applications. The battery voltage rises and falls with a large variation, as the engine speed and battery conditions change.

The two types of converters often found in boost–buck applications are known as SEPIC and Ćuk. These converters are similar, but the Ćuk converter has an inverted output, which means that the LED anode is connected to the ground rail. Like boost converters, overvoltage protection should be provided to prevent excessively high voltage in case of an open-load condition.

The Ćuk is good for low-EMI emissions because there are inductors in series with the input and the output. Both inductors operate in continuous conduction mode, and high-frequency signals at the central node, where the switching takes place, are automatically filtered. Shunt capacitors across the input and output strengthen this filtering, and provide a low-impedance path for the circulating currents. Consequently, Ćuk circuits require minimal external filtering. In the SEPIC circuit, the output inductor is connected to ground, so EMI is greater than for an equivalent Ćuk.

Sometimes common-mode chokes are added at the input side of a boost–buck converter, to reduce the radiated signals from the whole circuit. Common-mode chokes are only required on the output side if the length of wire to the LED load is more than about 0.5-m long.

10.2.4 Circuits With Power Factor Correction

Power factor is an indication of the relative phase of the power line voltage and the power line current. A power factor of 1 indicates that the voltage and current are in phase and have a low-harmonic content. A power factor of 0 indicates that the voltage and current are 90-degrees out of phase.

In semiconductor circuits powered from the AC mains, a bridge rectifier converts the AC power into DC. The current through the bridge rectifier tends to occur close to the peak voltage, as shown in Fig. 10.4 because charging of a large-smoothing capacitor takes place each half cycle. These short-charging current pulses at the crest of each input cycle cause the power factor to be typically in the 0.3–0.6 range. PFC is an active or passive circuit designed

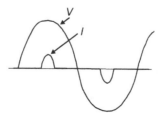

Figure 10.4: Active Circuit AC Input Current.

to correct phase errors and reduce harmonics, and make the power factor closer to 1. PFC is required in higher-power LED drivers.

A circuit having a good power factor, approaching 1, has an input current that has low-harmonic content with a wave shape that closely follows the sinusoidal input voltage. Circuits that provide a good power factor were described in Chapter 8.

10.2.5 Fly-Back Converter Considerations

Transformer-coupled switching regulators can be designed for a very wide range of supply and output voltages. The most common is a fly-back converter, although forward converters are also popular in higher-power applications. Fly-back converters were described in Chapter 9.

Fly-back converters allow an isolated LED driver design with about 80% efficiency, but have added cost and complexity. If a wide tolerance can be accepted for the current regulation, a simpler and cheaper circuit can be built. High accuracy requires isolated feedback, usually via an optocoupler and employing an adjustable shunt regulator, such as a TL431 or similar, along with a few passive components.

Fly-back converters have the advantage of stepping up or down the output voltage compared with the supply (buck–boost). This also applies to the single-winding inductor version, although as the same winding is used for the primary and secondary side, the turns ratio is 1:1 and the design specifications are more restricted than for a two-winding inductor. A single-winding inductor is usually much lower in cost.

A fly-back, by definition, is a discontinuous conduction mode converter; energy is taken from the power supply in the first step and then transferred to the output in the second step, as shown in Fig. 10.5. This means that EMI must be carefully filtered at both the input and output. The output requires a large-storage capacitor to maintain current flow in the LEDs when the converter is on the first step. Dimming the LED light by pulse width modulation (PWM) of the current is very difficult because the stored energy in the capacitor tries to maintain current flow; thus only a modest dimming range is possible.

Note that the LED current in this circuit is not controlled. Only the output power is controlled, so the LED current will depend on the LED string forward voltage, which is temperature dependent. Thus the LED current will change with LED temperature.

10.2.6 Inrush Limiters

As almost all circuits have decoupling capacitors, when a power source is connected there will be an inrush current. This current can be very high, causing temporary heating in the capacitor and possible damage to switch contacts or components connected in series. Inrush current limiting using passive or active components can be provided to reduce this risk.

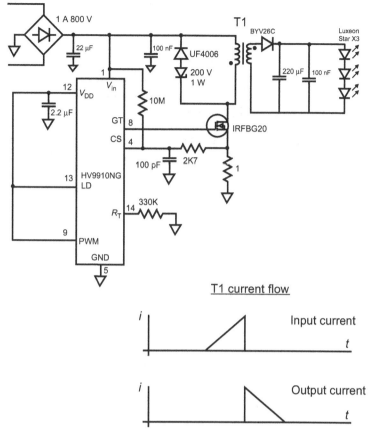

Figure 10.5: Discontinuous Fly-Back Current.
PWM, Pulse width modulation.

For AC mains applications, an NTC thermistor designed to carry high current is often used. In the active state, the flowing current warms the thermistor and hence the resistance falls to a low level to reduce losses (Fig. 10.6).

For DC applications, an active inrush limiter is more common. This is because the losses can be minimized during normal operation, when inrush limiting is not needed. There are several manufacturers of active inrush ICs, but Microchip's HV101 uses a unique scheme, whereby the Miller Effect is used to control the slow turn on of the MOSFET. An active inrush circuit using the HV101 is shown in Fig. 10.7.

10.2.7 Soft-Start Techniques

Some applications need the input current to be controlled, to prevent high-current spikes when power is first applied. This could be to reduce damage to switch contacts by the risk

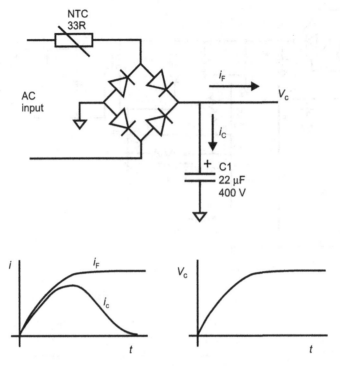

Figure 10.6: NTC Inrush Circuit.

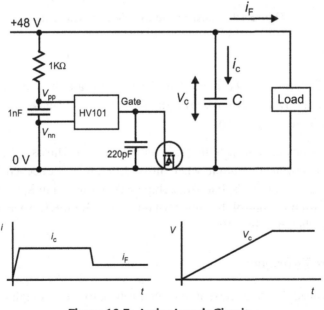

Figure 10.7: Active Inrush Circuit.

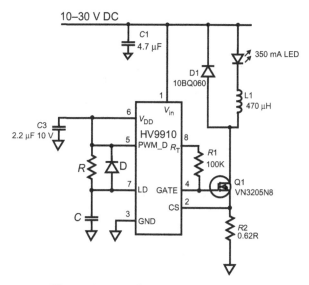

Figure 10.8: Soft-Start With HV9910.

of sparking. Clearly the inrush techniques just described could be used, but sometimes it is necessary to control the output power instead.

For example, a circuit for driving one or two power LEDs from the AC mains could use a double-buck topology. But typical applications for this circuit are inside lamp housings, where an electrolytic capacitor cannot be used because of short lifetime or physical size. But using a polyester film capacitor means that the voltage dips between switching cycles. As the output power is normally constant, this means that the input current will peak as the input voltage dips. The peaks in input current give rise to considerable EMI and mean that the power factor is very poor. If the output current was controlled, that is, reduced as the supply voltage dipped, the input current would remain constant when switching. The addition of a Zener diode in series with the supply to the controller IC would further improve the power factor.

Soft-start can also be implemented by connecting an RC filter to the analog dimming input in control circuits that have this function (e.g., linear dimming pin of HV9910). The current level starts low and grows as the capacitor, C, charges. Clearly we need a method of discharging this capacitor reasonably quickly when the power to the IC is disconnected; diode D connected to the V_{DD} will discharge capacitor C when the V_{DD} voltage drops (Fig. 10.8).

10.2.8 Slope Compensation

Slope compensation has been referred to in earlier chapters, but not described in detail because it is a topic for all types of switching regulators, whether buck, boost, or other types, and will be described here.

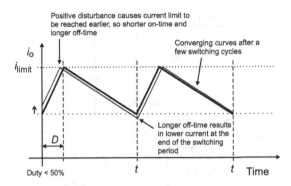

Figure 10.9: Switching Current With 25% Duty.

Slope compensation is needed if the following three conditions are met: (1) the switching is constant frequency, (2) the duty cycle is more than 50%, and (3) current mode control is used. Some DC/DC converters with a constant voltage output use voltage mode control; however, all LED drivers have a constant current output and use current mode control. In Chapter 5, I described buck controllers using constant off-time switching, to avoid the need for slope compensation. The disadvantage, in some applications, was that the switching frequency covered a wide range; so LED ripple current and EMI emissions were more difficult to control.

In Fig. 10.9, the switching inductor current is shown on a graph. In this case we are looking at a system with a duty cycle about 25%. The heavy black line shows the nominal current levels, with i_{limit} being the peak current level used to turn off the switch (peak level sensing). A thinner black line shows the effect of a disturbance, for example, a step positive-going increase in the supply voltage. Notice that the current limit is reached earlier, so the duty cycle is smaller, but the off-time is longer (as it is a fixed switching frequency) and the current drops below the original starting level.

On the second-switching cycle the current starts at a lower level than during the first-switching cycle, so the inductor current takes longer to reach the current limit. Now the on-time is longer, so the duty cycle is greater, and the off-time is shorter. At the end of the second cycle, the current is higher than the nominal level, but only by a small amount. On the following cycles, this difference dwindles to nothing, so the system is stable.

In Fig. 10.10, the graph shows the current in the switching inductor of a system where the duty cycle is 75%. As before, the heavy black line shows the nominal current levels, with i_{limit} being the peak current level used to turn off the switch (peak level sensing). A thinner black line shows the effect of a disturbance caused by a positive voltage step at the input. As before, the current limit is reached earlier. As the duty cycle is much higher in this case, there is a less steep slope on the graph and the disturbance causes a more significant reduction in the duty cycle. As the switching frequency is fixed, the off-time is increased significantly and the current drops to quite a low, way below the original starting level.

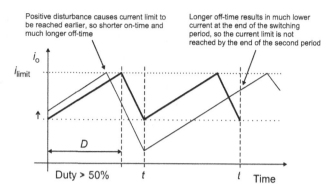

Figure 10.10: Switching Current With 75% Duty.

Thus the second cycle starts with the inductor current at a very low level and the current limit is not reached by the end of that switching cycle. Instead, the current limit is reached at some point in the next cycle. Thus the switch on-time will alternate between long and short periods, causing the circuit to operate at half the nominal switching frequency. This is sometimes called subharmonic oscillation. The ripple current will alternate between low and high levels. The average current will thus be much lower than expected and some engineers only realize that they have an instability problem when they measure the LED current.

In Fig. 10.11, a solution for unstable subharmonic oscillation is shown graphically. Instead of having a fixed current limit, the current limit is reduced over the switching period. At the start of each switching period, the current limit is maximal. At the end of each switching period, the current limit is at its minimum level. The slope of this current limit, shown by a thin line on the graph, is set to be half the downslope of the inductor current. This downslope is calculated at: Slope $= 0.5 \times V_{out}/L$. The maximum slope is Slope$_{max} = 0.5 \times V_{out\,max}/L_{min}$, which should be used in calculations.

Now, as in the previous examples, the thick black line shows the nominal current and the thinner black line shows the effect of a disturbance, such as a step increase in the supply

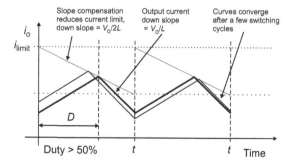

Figure 10.11: Slope Compensation Added to Current Limit.

voltage. After the input voltage disturbance, the current limit is reached earlier than with the nominal input voltage, but the sloping current limit has the effect of reducing the difference between the two levels by the end of the first-switching cycle.

At the start of the second-switching cycle, the inductor current starts at slightly below the nominal level. The difference is much less than at the beginning. Due to the sloping current limit, the difference between the two levels is much smaller after the second-switching cycle and becomes insignificant on subsequent cycles.

In practice, the slope compensation is arranged by adding a voltage to the current sense feedback signal. When the current sense voltage and the added slope compensation voltage reach a threshold level, the power switch is turned off. The switch will not be turned on again until the clock signal triggers the gate drive output from the LED driver IC, so the slope compensation ramp signal is only needed while the power switch is on. Thus a ramp signal can be created by charging a capacitor from the power switch gate drive signal. In some cases, an LED driver IC will include a constant current circuit driven from the gate output signal, so the user only has to add an external capacitor connected from a COMP pin to ground.

Now, with slope compensation added, we can see that the current limit is now lower than before. To keep the desired (original) current levels, the current sense resistor value now has to be scaled down in proportion.

$$\text{Downslope} = 0.5 \times V_{\text{out max}} / L$$

$$\text{Period} = T, \text{Duty} = D, \text{so on time} = D \times T.$$

$$\text{Reduction in } i_{\text{out max}}, \Delta_{i_{\text{out}}} = \frac{0.5 \cdot V_{\text{out max}} \cdot D \cdot T}{L}$$

$$R_{\text{sense new}} = R_{\text{sense old}} \cdot \left| \frac{i_{\text{out max}} - \Delta_{i_{\text{out}}}}{i_{\text{out max}}} \right|$$

The new value for R_{sense} can now be used with slope compensation, to give the correct (higher) current limit.

CHAPTER 11

Selecting Components for LED Drivers

Chapter Outline

This chapter will be very practical in orientation. It will describe how different materials and component types can affect the performance of LED drivers. This will be detailed, showing how the physical construction of components could have an effect.

11.1 Discrete Semiconductors

Atoms of materials have a core (nucleus) of positively charged proton and uncharged neutrons. They have negatively charged electrons orbiting around this nucleus, like planets around the Sun. When atoms combine, they share electrons in their outer orbit (the valence band). Lighter atoms, like silicon, are most stable when there are eight electrons in their outer orbit. Semiconductors are (usually) made from silicon, which has four electrons in its outer orbit.

The addition of a small amount of material (dopant) with either three or five electrons in their atom's outer orbit can create an imbalance because, when combined with the four electrons of silicon, there are either seven or nine electrons in the outer orbit. When doped with material having three electrons in the valence band [boron (B), aluminum (Al), gallium (Ga), or indium (In)], the resultant outer orbit has seven electrons and a "hole" where an electron is

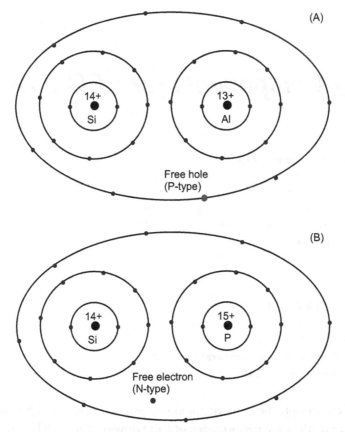

Figure 11.1: (A) P-Type atomic level and (B) N-Type atomic level.

missing. This hole appears as a free positive charge and is called P-type semiconductor. This is shown in Fig. 11.1A.

When doped with material having five electrons in the valence band [phosphorous (P), arsenic (As), or antimony (Sb)], the resultant outer orbit has nine electrons which means that there is a "free" negatively charged electron and the material is called N-type semiconductor. This is shown in Fig. 11.1B.

When P-type and N-type semiconductor form a junction, the free electrons and holes combine and are destroyed. The fixed nuclei have a net negative and positive charge, respectively, and thus repel the combination of further free electrons and holes. Thus there is an energy barrier created; we have a diode junction. This is shown in Fig. 11.2.

In order for a P–N junction to conduct, we must make the P-type material more positive than the N-type. This forces more positive charge into the P-type material and more negative charge into the N-type material. Conduction takes place when (in silicon) there is about 0.7 V potential difference across the P–N junction. This potential difference gives electrons enough energy to conduct.

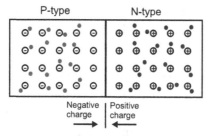

Figure 11.2: P–N Junction Diode.

11.1.1 MOSFETs

Metal oxide silicon field effect transistors (MOSFETs) are used as electronic switches in switching and linear LED driver circuits. They operate by using the "field effect" in semiconductors; where an electric field attracts or repels free electrons in doped silicon. A MOSFET has three terminals—gate, drain, and source; a fourth "body" terminal is internally connected to the source. A diagram showing the physical construction of the MOSFET is shown in Fig. 11.3.

Notice that the source and body are connected together by the metalized contact at the source. Also notice that there is a parasitic diode due to the P-type material of the body and the N-type material of the drain. This parasitic diode is reverse biased normally because the drain is more positive than the body (and source), so does not need to be considered

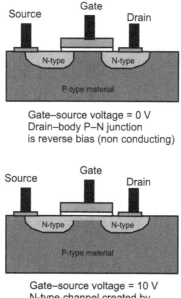

Figure 11.3: N-Channel MOSFET Construction.

in all applications. Note that the body diode generally has a long reverse recovery time; in synchronous switching applications where the body diode may conduct, an ultrafast diode is sometimes connected in parallel to prevent or reduce the body diode current.

To create a conducting channel in the body of the MOSFET requires a certain amount of gate potential. MOSFETs are specified with a certain gate threshold voltage, usually at the point where the drain current reaches 1 mA, but this varies between manufacturers. As the gate–body isolation is a dielectric, gate–source and gate–drain capacitance values are usually found in the datasheet.

Typical gate thresholds are in the range 4–7 V, however a number of "logic level" devices are now available. A "logic level" device is defined as one that switches fully on at V_{gs} equal to 5 V; this means that the gate threshold is typically about 2 V. So-called "standard devices" are defined as being fully switched on at V_{gs} equal to 10 V. A logic level device can also be operated with V_{gs} equal to 10 V or higher, in which case the on-resistance is lower. Many logic level devices have a higher gate capacitance compared to standard devices, for a comparable saturation current rating.

Fig. 11.4A shows a simple MOSFET circuit with parasitic capacitance C_{gd} and C_{gs}; Fig. 11.4B shows a graph of gate voltage versus gate charge. This helps to understand the switch operation when the MOSFET is driven from a current source because the rate of charge is constant (Q_g = current × time) and thus the horizontal axis represents time. The gate voltage initially rises quickly as C_{gs} is charged. Once the gate threshold voltage, $V_{gs(th)}$, is reached, the MOSFET starts to turn on; this causes the drain voltage to fall forcing the gate drive current into capacitor C_{gd} instead of C_{gs}, so the gate voltage rises very slowly while the drain voltage falls.

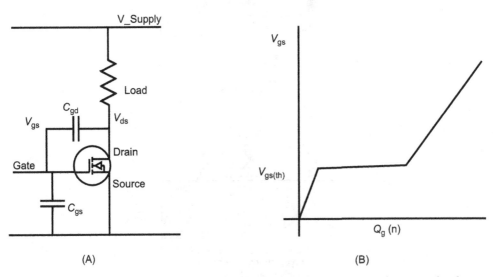

(A)   (B)

Figure 11.4: (A) MOSFET circuit with parasitic capacitance shown and (B) graph of gate charge vs. gate voltage.

Once the MOSFET is turned on and the drain voltage is close to 0 V, the capacitors C_{gd} and C_{gs} are both charged at a steady rate. Note that both C_{gd} and C_{gs} are capacitors formed across a reverse bias P–N junction and so will have lower capacitance when the reverse bias voltage is high. Now that the drain voltage is close to 0 V, the capacitances of both C_{gs} and C_{gd} have increased because of the small reverse bias voltage, and the charging rate is lower, thus the slope of the curve is now less steep than during the initial charge (below $V_{gs(th)}$).

Note that a MOSFET gate is normally driven from a resistive or current-limited source. In fact designers often add a resistor in series with the gate, for two reasons: (1) to reduce EMI, by slowing down the turn-on speed; and (2) to prevent high frequency ringing at the MOSFET drain due to resonance of the drain–gate capacitance and the load inductance. The load inductance may be from the inductor in a buck converter, or a transformer primary in a fly-back converter.

MOSFETs have two current ratings—peak current and continuous current. Continuous current ratings depend on the on-resistance of the MOSFET and are based purely on thermal considerations. Peak current ratings are the maximum current that is able to flow. When designing a switching LED driver circuit, the circuit current is pulsed and so the peak current rating is important. However, note that this current is normally quoted at 25°C; at 100°C the peak current is about half this value. As a rule of thumb, always use a MOSFET that has a peak current rating that is 3 times the value needed in the application.

11.1.1.1 Miller Effect

When the MOSFET is connected to a load, but turned off, the drain is at high voltage. When the gate voltage rises, the MOSFET turns on and the drain voltage falls close to the ground (0 V) potential. The gate–drain capacitance thus sees a large voltage fall on the drain side and a slight rise on the gate side. At the gate pin, the gate–drain capacitance appears to be much larger than it really is; this is known as the Miller effect, named after the engineer who discovered this phenomenon. The parasitic capacitance C_{gd} and C_{gs} in a simple MOSFET circuit and the gate charge graph were shown earlier in Fig. 11.4.

Two MOSFETs are sometimes connected in parallel to carry high switching current. If the gates are connected directly together, there could be a problem. This is because each MOSFET is likely to have a different gate threshold voltage and therefore one will start to turn on before the other. Suppose that MOSFET M1 has a 4 V threshold and MOSFET M2 has a 5 V threshold. When the gate drive circuit's output voltage rises to about 4 V, M1 will start to turn on and because of the Miller effect, the gate voltage will be held to just above 4 V until the gate–source capacitance C_{gs} is fully charged. During the C_{gs} charging time, M1 will carry all the current because M2 has insufficient gate voltage to turn on.

To drive two parallel MOSFETs properly, a resistor should be connected in series with each gate. The gate driver output voltage will rise to the maximum level quickly because the gate resistors prevent the gate capacitance from holding the voltage down. Each gate will charge separately and the difference between turn-on times will be minimized.

11.1.1.2 Gate Charge

Instead of considering the gate–drain capacitance and the gate–source capacitance, we can consider the gate charge. This is the total charge needed to turn the MOSFET on. In switching circuits, the gate charge is most significant and is usually quoted in nanocoulombs (nC). The average gate current is given by the equation:

$$I_g = Q_g \cdot F_{sw}$$

The current drain from the supply to power the gate driver will be equal to this average gate current. Therefore the average current to power the LED driver IC will be a small quiescent current for internal functions, such as the oscillator and voltage references, plus the product of gate charge and switching frequency.

$$I = I_q + Q_g \cdot F_{sw}$$

This is important when calculating the power dissipation in a MOSFET driver circuit. The power dissipation will be the product of supply voltage and IC current, calculated earlier.

11.1.2 Bipolar Transistors

Bipolar transistors are used in switching and linear LED driver circuits. They operate by a current magnification effect; the collector–emitter current is a multiple of the base–emitter current. The base–emitter voltage is about 0.7 V, being the voltage drop of a forward biased P–N junction. There is some base–emitter resistance, so the forward voltage drop will increase slightly with base current.

Matched bipolar transistors can be very useful, particularly in current mirror circuits. A current mirror is one where two or more branches carry identical currents; the current in one branch depends on the current in another, hence the "mirror." Transistors do not have to be matched to make a current mirror. Transistors of the same type have very similar characteristics, so by adding a low value resistor between the emitter and ground any variation in the base–emitter voltage (V_{be}) is negligible; see Fig. 11.5.

11.1.3 Diodes

There are many different diodes (rectifiers). Important parameters include: reverse breakdown voltage, forward current rating (average and peak), forward voltage drop, reverse recovery time, and reverse leakage current.

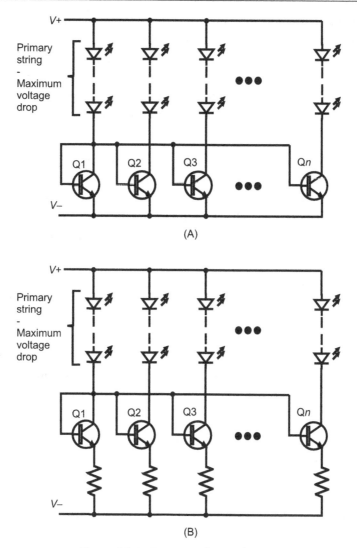

Figure 11.5: Current Mirror Circuits.
(A) Matched NPN and (B) nonmatched NPN.

Schottky diodes have the lowest forward voltage drop and the shortest reverse recovery time, but they are more expensive than standard diodes and generally have a limited reverse breakdown voltage range. Instead of a P- and N-type semiconductor junction, the Schottky diode has an N-type semiconductor and metal junction. Reverse leakage is higher than in most P–N junction diodes. They are used for many applications, including reverse polarity protection and as flywheel diodes in low voltage switching circuits. Note that the forward voltage drop across a Schottky junction tends to increase with diode voltage rating, so use the lowest voltage rating suitable to keep the conduction losses to a minimum.

Diodes are sometimes labeled by their reverse recovery time. When the voltage across a diode is suddenly reversed, an initial current flow will occur in the reverse direction. Reverse recovery time (T_{rr}) is the time taken to stop conducting when the diode is reverse biased. The labels, fast, ultrafast, and hyperfast are sometimes given. A standard rectifier diode like 1N4007 has a typical reverse recovery time of 30 µs, but an ultrafast version UF4007 has $T_{rr} = 75$ ns, which is about 500 times faster. More recent devices are much faster, for example, the STTH1R06 is a 600 V, 1 A rectifier with $T_{rr} \sim 30$ ns.

There are now high voltage Schottky diodes available from a few suppliers, notably Cree and ST Microelectronics. These are known as silicon carbide (SiC) diodes and are rated from 600 V to above 1 kV. These type of diodes are very useful in circuits like power factor correction boost circuits and high voltage, high current, buck circuits, where the short reverse recovery time prevent high switching losses. They tend to be more expensive than ultrafast junction diodes.

Shorter reverse recovery times reduce switching losses. This is because the reverse current often flows through the MOSFET switch when the voltage across the MOSFET is high, so a shorter time gives lower losses. However, a "snappy" diode can sometimes generate radio interference (EMI), due to the fast turn-off causing high frequency ringing. In some applications a "soft-recovery" diode should be used, where the turn off speed in the reverse-biased condition is fast but at a controlled rate of change. If slower diodes must be used, slowing the MOSFET switching speed by adding a resistor in series with the gate may be necessary to prevent the diode from overheating.

In fly-back power supplies, a series connected RC snubber circuit is placed across the primary winding to prevent very high voltages when the MOSFET switch turns off. Some snubber circuits use a medium speed diode in series with a resistor, so that the diode blocks current during switch turn-on, but is conducting in both directions for a period of about 250 ns after switch turn-off. This allows any ringing current to flow through the resistor and thus decay quickly. Alternatively, a fast diode connected in series with a high voltage Zener diode (the two anodes connected together) can be used to clamp ringing.

11.1.4 Voltage Clamping Devices

Voltage clamping devices are used to limit the voltage across a circuit, as part of a voltage regulator or a transient suppressor. These devices are typically semiconductors: Zener diodes, Transorb suppressors, or voltage-dependent resistors (VDRs).

Zener diodes behave like regular diodes in the forward conducting direction, but breakdown and conduct at a defined voltage in the reverse direction. Low voltage Zener diodes rated below 6 V have a soft knee in their current versus voltage graph; the conduction increases gradually. High voltage Zener diodes (avalanche diodes) rated above about 6 V have a sharp knee and conduction increases very rapidly. Zener diodes can exhibit some noise when breaking down and are often used with a small capacitor in parallel to reduce this effect.

Transorb suppressors are like Zener diodes but are designed to handle high current peaks. Transorbs can be unidirectional or bidirectional and rated from low voltage ~5 V up to several hundred volts. A Transorb designed for 275 V AC operation will limit the peak surge voltage to below 600 V, even at high transient current levels.

A VDR has high resistance at low voltage and low resistance at high voltage. Thus conduction increases gradually as the voltage across it increases. A VDR can absorb high surge energy; the devices are often rated in Joules rather than Watts because the surge energy is short lived. A VDR rated at 275 V AC will breakdown and limit the voltage to about 710 V at high transient current levels.

11.2 Passive Components

Passive components are capacitors, inductors, and resistors. These may appear to be simple, but often parasitic components are also present and we have to consider what effect they will have on the circuit behavior.

11.2.1 Capacitors

Capacitors are constructed from two conducting surfaces (known as plates) separated by an insulator (known as a dielectric). The metal plates are made from a thin metal film that has been deposited onto the insulation material. The dielectric can be a number of materials including ceramic, mica, and plastic film. The capacitor type is usually known by the dielectric, thus there are "(aluminum) electrolytic" capacitors, "ceramic" capacitors, "polyester" capacitors, and so on.

A capacitor's behavior is not ideal. Every capacitor will have some series inductance; this is due to the plate conductors and the lead wires attached to them. This self-inductance can be a problem at frequencies close to the self-resonant frequency and above. Each capacitor will also have series resistance, due to both the conductors and the dielectric of the insulator, this is known as equivalent series resistance (ESR). The ESR will create losses. An equivalent circuit for a capacitor is shown in Fig. 11.6.

In an LED driver, the key function of a capacitor (symbol C) is energy storage. There are two types of storage, slow storage and fast storage. Slow storage is required across the DC

Figure 11.6: Capacitor Equivalent Circuit.

supply, to hold up the voltage when powered from a low frequency AC supply. Sometime direct connection is possible (for very low power lamps), but more often a power factor correction stage produces a high voltage DC energy that must be stored. The purpose of this energy storage is to supply power to the LED driver between the peaks of the AC voltage, which is twice every cycle. The AC frequency is typically 50–60 Hz, although 400 Hz is used in some aircraft, so the capacitor must supply energy and hold up the supply for as long as 10 ms.

For slow storage, an aluminum electrolytic capacitor is often used because it has a high energy storage density (they take up less space for an equivalent amount of storage, compared to other dielectric types). These capacitors are made using aluminum foil with a wet dielectric material. Due to this construction, they cannot be used in a high temperature environment for long periods; the dielectric dries out and the capacitor eventually fails.

Good quality electrolytic capacitors may have a lifetime of 10,000 h at 105°C, for example. But electrolytic capacitors do not suddenly fail, they decay slowly. Lifetime is defined by the manufacturer: some use the value of ESR, where lifetime is when the ESR has doubled from its initial value. Doubling of the ESR means that power dissipation will be doubled for a certain level of ripple current, so some manufacturers measure the maximum ripple current for a given temperature rise in the capacitor. An important "rule of thumb" is that lifetime doubles for every 10°C drop in ambient temperature, so operation at 85°C, giving a 20°C drop, would give 4 times the lifetime (40,000 h for a 10,000 h 105°C rated capacitor). This means 5 years lifetime.

Fast energy storage is required in switching driver circuits, where the switching frequency is often in the range 50–500 kHz. The energy only has to be stored for a short time, as short as a few microseconds, so the main characteristic of the capacitor for this function is to have the ability to store and discharge energy quickly. This means low self-inductance (high self-resonant frequency). Surface-mount components generally have lower self-inductance because they have no added lead inductance. Typically, the capacitors used for fast energy storage are ceramic or plastic film types.

Ceramic and mica capacitors are made using flat dielectric sheets; the simplest construction uses just one insulating layer with a conducting plate on either side. Mica capacitors are very rarely used, but ceramic are fairly common. Higher valued devices use several insulating layers with interleaving layers of metal film. The metal film layers are bonded alternatively to side A, side B, side A, side B, etc.

Plastic film capacitors, such as polyester, polypropylene, polycarbonate, and so on, use two layers of metalized plastic film. One form of construction is identical to that of ceramic capacitors, where flat sheets of metalized film are used. This type of construction is often found in surface-mount polyester capacitors.

Another form of construction for plastic film capacitors uses rolled films. Two metalized layers are placed one above the other and then rolled, so that the two conductors spiral around each other with insulating layers in between. The films are laterally offset from one another so that the conductor of "side A" protrudes from one side, and the conductor of "side B" protrudes from the other side (this technique is sometimes known as extended foil). It is then relatively easy to bond lead-wires to the ends of the resulting cylindrical body. The rolled form of construction provides a metal film around the body of the capacitor; this can be connected to earth or the "earthy" side of a circuit to reduce external electric field pickup. Connecting the outer foil to earth also helps to reduce EMI. The outer foil is marked on the outer case of some film capacitors.

Generally, ESR and self-inductance is more of a problem with aluminum or tantalum electrolytic capacitors. These types of capacitors are normally used to decouple power supplies. Digital circuit designers have long since become accustomed to connecting 10 nF ceramic capacitors across tantalum devices used for power supply decoupling. This is because the higher value tantalum capacitor absorbs low frequency transient currents, while the ceramic absorbs the high frequency transient currents.

Dissipation factor (DF) and loss tangent are terms used to describe the effect of ESR. The value of DF is given by the equation:

$$\text{Loss tangent} = \text{DF} = \frac{\text{ESR}}{X_c},$$

where X_c is the capacitor's reactance at some specific frequency. This is the tangent of the angle between the reactance vector X_c, and the impedance vector $(X_c + \text{ESR})$, where the ESR vector is at right angles to the reactance vector.

One of the most notable problems with capacitors is self-resonance. Self-resonance occurs due to the parasitic inductance mentioned earlier. Consider the self-resonant frequency of capacitors, of various dielectrics, having a lead length of 2.5 mm (or 0.1 in.): a 10 nF disc ceramic has a self-resonance of about 20 MHz; the same value of polyester or polycarbonate capacitor also has a self-resonance of about 20 MHz.

A rough idea of the self-resonant frequency can be found by calculating the inductance of a component lead. For example, a 0.5-mm diameter lead that is 5-mm long (2.5 mm for each end of the component) has an inductance of 2.94 nH in free space. When combined with a 1-nF capacitor, the self-resonant frequency is calculated to be about 93 MHz. Replacing the 1-nF capacitor in previous calculations with a 10-nF capacitor, results in the self-resonant frequency falling to 29 MHz.

Earlier, I wrote that the self-resonant frequency of a 10-nF capacitor with 2.5-mm leads was about 20 MHz, not 29 MHz. The reason for the discrepancy between the calculated frequency and the actual frequency is that inductance in the plates was not taken into account. Adding the inductance of the plates gives a lower self-resonant frequency. As the value of the capacitor increases, the inductance of its plates also increases and so does the discrepancy between the calculated and the actual self-resonant frequency.

For small value capacitors of less than 1 nF the self-resonant frequency can be approximately calculated by the following equations.

$$f_R = \frac{1}{2\pi\sqrt{LC}},$$

where L is the lead inductance. For a wire in free space, $L = 0.0002\,b\left\{\left[\ln\left(\frac{2b}{a}\right)\right]-0.75\right\}\mu H,$

where "a" equals the lead radius, b equals the lead length. All dimensions are in millimeters (mm) and the inductance is in H.

Using the formulae, if $a = 0.25$ mm (0.5-mm diameter) and $b = 5$ mm (2.5-mm each leg), the inductance is $2.94 \times 10{-}3$ μH. This is 2.94 nH. When substituted into the frequency equation, with a 1-nF capacitor, the self-resonant frequency is calculated to be 92.8 MHz.

Surface-mount capacitors are in common use now because of their small size. In the past they were often used for high frequency circuits because there is no lead inductance to worry about. This reduction in inductance has benefits for switching power supplies too; where fast pulse rise and fall times are needed. The most popular type of surface-mount capacitor is the multilayer ceramic; it has conducting plates that are planar and interleaved; they have very little parasitic inductance. Some conventional leaded ceramic capacitors are surface-mount devices with wire leads attached. They are usually dipped in epoxy resin or similar before having their capacitance value and voltage rating marked on the outside.

Ceramic capacitors generally have a temperature coefficient that is zero or negative. The terms NPO or COG are used to describe ceramic capacitors with a zero temperature coefficient (NPO = negative positive zero). Other ceramic dielectrics are described by the temperature coefficient; N750 describes a dielectric that has a negative temperature coefficient of -750 ppm/C. More exotic dielectrics are X7R, X5R, and Y5U, which have a higher dielectric coefficient and are used to make capacitors with high capacitance values.

The X7R and Y5U capacitors have a wide tolerance on the component value. The first character defines the minimum temperature, so $X = -55°C$ and $Y = -30°C$, for example. The second character defines the upper temperature limit, so $5 = 85°C$ and $7 = 125°C$. The third character defines the capacitance change over temperature, so $R = \pm15\%$ and $U = +22\%$, -56%. Perhaps Y5U should be avoided.

Even if we keep to X7R and X5R capacitors, we have another characteristic to consider. The capacitance of a ceramic capacitor changes with applied voltage. Also, the smaller the

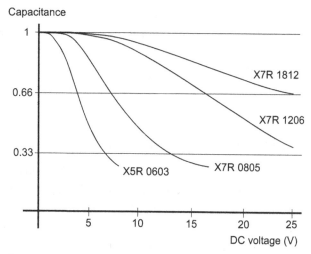

Figure 11.7: Capacitance Change With Applied Voltage.

package size, the worst this effect becomes. For example, a 0603 capacitor may have its original (0 V bias) capacitance value reduced to less than half with just 6 V bias, so a 2.2-µF capacitor looks like 1 µF. This is illustrated in Fig. 11.7.

This reduction in capacitance with applied voltage is important in many designs. Consider a decoupling capacitor used for the internal power supply of an LED driver, it must supply high current pulses at high frequency to drive the gate of the power switching MOSFET. Calculations may have shown that a 1-µF capacitor would be sufficient, except that we discover that the effective capacitance has been reduced to 400 nF when a DC bias was applied. A capacitor of the same value, but in a much larger 1210 package, would show perhaps 10% drop in capacitance with 6 V bias. Unfortunately, size and cost can be critical in some applications and a 1210 size package may not be an option.

Apart from NPO/COG capacitors, ceramic capacitors exhibit a piezoelectric effect. A high voltage AC signal can generate acoustic noise. The acoustic output increases with physical size, so a surface-mount 1206 size capacitor will generate more noise than a 0805 capacitor in the same circuit. The piezoelectric effect will also cause the capacitance value to change with applied voltage, as described earlier. In applications like alarm clocks and theater lighting, any noise is a big problem because of the very quiet environment they operate in, so alternative capacitor types are recommended.

Polystyrene and polypropylene capacitors have a negative temperature coefficient that fortunately closely matches the positive temperature coefficient of a ferrite-cored inductor. They are thus ideal for making LC filters. Unfortunately, with these dielectrics, capacitors tend to be physically large for a given capacitance value.

Polyester and polycarbonate capacitors are commonly used in LED driver circuits. Polyester capacitors have the worst performance in that they have a poor power factor (high ESR) and a poor (and positive) temperature coefficient. But polyester capacitors are popular because they have a high capacitance density (high capacitance value devices are small) and lower cost. Polycarbonate capacitors have a better power factor and a slightly positive temperature coefficient. Another useful feature of polycarbonate capacitors is that they are "self-healing"; in the event of an insulation breakdown due to overvoltage stress, the device will return to its nonconducting state, rather than become short circuit.

Capacitors used across the AC mains supply must be rated X2. For universal AC input, 275 V AC X2 rating is normally used. These capacitors are available in polyester and polypropylene, and 100 nF is a typical value found across the supply connections. This capacitor reduces EMI emissions and absorbs fast transient surges from the mains supply. In a typical application, a VDR (or varistor) is connected in parallel.

In some applications, a standby power supply is created by having a series capacitor (usually X2 rated) and a shunt Zener diode (or an active shunt like the Microchip SR10). One problem with series capacitors is that the capacitance value falls over time. Corona discharge during overvoltage conditions (line borne surges) causes the metalization layers to be burnt away, thus reducing the capacitance value. The current available for the power supply will reduce over time. An example that I have seen is an electric shower control unit that failed because the series-connected 100-nF capacitor, used in its power supply, had its capacitance value reduced to 25 nF after about 3 years of service. It is possible to obtain "low corona" capacitors, which have a longer lifetime in such circuits, but they tend to cost more.

Capacitors from each AC power line to ground (earth) are sometimes used and must be 250 V AC Y2 rated. These capacitors typically have a dielectric of ceramic, polyester, or polypropylene. Capacitance values are readily available in the range 1–47 nF. A value of 2.2 nF is commonly found in power supply designs.

11.2.2 Inductors

This section will describe "off-the-shelf" inductors and transformers. Details of custom-made components will be covered in Chapter 12.

Inductors (symbol L) are used to store energy in switching LED driver circuits. A length of wire creates inductance, but winding insulated wire into a coil can magnify this effect because the magnetic field produced by a wire then couples to adjacent wire. With intercoil coupling, the total inductance is proportional to the number of turns squared. The wire used in multiturn coils is normally soft copper covered with a thin plastic film for insulation; this is sometimes referred to as enameled copper wire (ECW).

Although a simple coil creates inductance, if a magnetic material is placed within the coil the inductance increases considerably. The coil can be wound around a short ferrite or iron-dust rod to increase their inductance, but with this type of core a magnetic field will radiate and may cause interference (EMI). The advantage of this type of inductor is that the saturation current level is very high, considering the inductor size. A typical application for this type of inductor is in the power filter at the input of an LED driver. Many low value inductors have the same appearance as wire-ended resistors, with colored bands marking their inductance value.

Alternatively, the coil can be wound around a toroidal (doughnut) shaped ferrite or iron-dust core. The toroidal shape keeps the magnetic field contained. Some toroidal materials have a distributed air gap, in which case the saturation current is very high. Toroidal inductor winding is not easy, requiring specialized winding equipment, hence these types of inductors can be more expensive than their bobbin-wound counterparts.

Shielded bobbin cores are popular. The coil is wound on a bobbin, within a closed ferrite material. These are of low cost and small physical size. Many can be surface mounted, which make printed circuit board (PCB) assembly easier. The central ferrite core inside the coil often has an air gap to increase the saturation current rating, although this reduces the inductance value.

An inductor's behavior is to oppose any change in the current flowing through it. This is because the energy stored in an inductor is given by $E = \frac{1}{2}LI^2$. To change the current instantly through an inductor would take infinite power. If we ignore physical imperfections due to the construction of an inductor, when a voltage is applied across it the current will increase linearly. If a load is then applied across the inductor, the current falls linearly. If we alternately switch the voltage source and the load across the inductor, the current will rise and fall, but remain fairly constant.

Inductors can be used to filter the power supply lines in switching LED drivers. Due to their energy storage characteristics, they tend to oppose any change in current, so they present high impedance to unwanted interference. Combined with capacitors that are low impedance to unwanted interference, the resulting "T" or "PI" filter considerably reduces the amplitude of high frequency signals.

Inductors can be a source of many problems. High value inductors are bulky. This is because they are usually made up from tens or hundreds of turns of ECW that is wound on a ferrite core. The windings capacitively couple to each other, which effectively introduces a parallel capacitor across the coil. This capacitance causes switching losses in power supplies, or poor filtering in supply input filters. Above the self-resonant frequency, the impedance of the inductor falls due to the dominating capacitive reactance.

Inductors also possess some series resistance due to the intrinsic resistance of the copper wire used. This resistance will cause losses in the power supply and thus limit the efficiency. Heating effects due to this resistance can cause problems. Choosing an inductor for the correct inductance value, without considering the ESR and self-resonant frequency will give poor results.

Magnetizing (core) losses are also present and are due to the energy required to make the magnetic fields in the core to align with each other. In a switching circuit these losses are continuous and can cause core heating. These losses increase rapidly if the magnetization is forced to operate outside its linear region. The presence of an air gap in inductor and transformer cores makes them suitable for high magnetic saturation levels. Transformer cores that have no air gap are prone to saturate easily.

The saturation current (I_{sat}) quoted by manufacturers is usually at the point where the inductance drops by 10%. If the inductor is expected to handle discontinuous switching current, where the current drops to or near zero each switching cycle, the peak current should be kept well below the saturation level (I suggest $I_{max} = 0.5 \cdot I_{sat}$, but preferably $0.25 \cdot I_{sat}$). This is because magnetization losses will cause core heating. Take care, because sometimes the current rating given by manufacturers is the DC current that causes a certain amount of heating, due to the winding resistance; the saturation current could be a lower current value. Some manufacturers quote a saturation current at the point where the inductance has fallen to 60% of the zero current value!

Sometimes an inductor data sheet will give a "Q" value at a certain frequency. This is the voltage or current magnification value in a tuned circuit. It indicates the ESR of an inductor, $Q = \dfrac{\omega L}{R}$, which is more accurate than the DC resistance measurement. This is because of the "skin effect."

The "skin effect" raises the resistance of wire at high frequencies. This effect occurs when AC current flows in the wire, producing a magnetic field concentrated at the center of the wire. This magnetic field forces the electrons to travel down the outside surface (hence "skin" effect). The effect gets stronger as the AC frequency increases. This can be a serious problem for inductors working at a few hundred kilohertz, but can be alleviated by the use of multiple stands of insulated copper wire, twisted together.

Litz wire is made from a number of ECW strands, with an overall covering of cotton braid. This is the type of wire used to make ferrite rod antennas for radios working in the low and medium frequency range (LF and MF). This wire has a lower skin effect because the current is shared down each of the strands; the surface area of all the strands combined is considerably larger than the equivalent diameter of solid copper wire.

More recently, copper wire suppliers have been producing flat wire, which is a thin strip rather than a round wire. Flat wire has a much larger surface area than a round wire of similar cross-section and thus exhibits less skin effect at high frequency. Flat wires are suitable for use in inductors and transformers, in high frequency switching power supplies

Off-the-shelf transformers are available with double or multiple windings, with or without an air gap in the magnetic core. Fly-back power supplies, including isolated LED driver circuits, use gapped cores; the air gap allows high magnetic flux density within the core—the energy is stored and then released. A forward converter is a popular power supply topology, which uses ungapped cores because magnetic energy is not stored in the core—it is immediately transferred to the secondary winding. Forward converter power supplies are rarely used in LED driving.

Transformers with multiple windings are used to create a step-up or step-down primary-to-secondary turns ratio. This allows the duty cycle of the switching circuit to set within a certain range. Very small duty cycles less than 5% should be avoided because of the difficulty in controlling the switching (due to delays in the system). Duty cycles greater than 50% can cause instability unless external compensation circuits or slope compensation are added. In some cases, such as where the input voltage range is very wide, a wide range of duty cycle may be unavoidable.

Another reason for an additional winding is to create a "bootstrap." A bootstrap circuit creates a power supply for the switching circuit, typically in the range 8–15 V. The switching circuit will be powered from the main power source initially, but this can be inefficient if the power source is of high voltage. Once switching starts, the voltage developed on the bootstrap winding can be used to self-power the switching circuit. Suppose the device needs 2 mA to operate, when powered from a 300 V DC supply it will dissipate 600 mW, but when powered from a bootstrap winding at, say, 10 V it only dissipates 20 mW.

11.2.3 Resistors

There are several types of resistors. Wire-wound devices are rarely used and would not normally be placed in a circuit that carried high switching current because they have a high self-inductance. They are used at the AC power input of some power supplies to provide some impedance for fast transients and surges. Carbon composition resistors tend to be noisy and have a poor temperature coefficient, but are good in switching power circuits because of their low inductance construction. They are constructed using carbon particles set in a clay rod and the resistance depends on the surface area of the touching particles. Carbon film and metal film devices are most common; surface-mount film devices are usually of thick film construction.

Carbon film resistors are low noise devices with a negative temperature coefficient. Component tolerances of 1 and 5% are standard, although 0.1% are available albeit more expensive. Through-hole resistors are constructed by applying a carbon film onto a ceramic rod, and then cutting a spiral gap in the film to increase the resistance. The spiral conductor is actually a lossy inductor. Surface-mount devices have a carbon film applied to one side of a ceramic sheet and a laser is used to cut part way across the film to alter the resistance. The short length of carbon film has very little inductance.

Metal film resistors have a lower noise than carbon film types, and a lower temperature coefficient. Component tolerances of 1% are standard, although precision devices in an E96 range of values with 0.1% tolerance and 15 ppm temperature coefficient are available at a higher cost. These resistors are constructed by applying a number of metal film layers, of different metals, to a ceramic former to achieve the correct resistance and a low temperature coefficient. In through-hole resistors, a spiral gap is sometime cut around the metal film to increase the resistance value and this increases the inductance slightly.

All conductors have some series inductance, simply due to having a certain length. This is typically 6 nH/cm. In fact some high frequency circuits just use a thin wire bond to form an inductor. Resistors are conductors and therefore have inductance too. Some types have more inductance than others. Even a thick-film surface-mount resistor has inductance, although of considerably lower value than other types. Wire-wound resistors have a significant inductance because of their construction; as we have seen with inductors, when a wire is wound into a coil its inductance increases in proportion to the number of turns squared. Carbon or metal film resistors that have had a spiral gap cut through their surface will have more inductance than a carbon composition type. All through-hole components will also have some inductance due to the wire leads connected at either end.

Resistors also have capacitance. The two ends have a certain cross-sectional area and are spaced a certain distance apart, separated by a ceramic dielectric. This capacitance is small, typically 0.2 pF, so has little effect in an LED driver circuit operating up to 1 MHz. However, at high frequencies this parasitic capacitance can be significant in parts of the circuit that are of high impedance.

11.3 The Printed Circuit Board

The circuit board on which the components are connected is important at high frequencies and for surface-mount circuits. There are several types of board, with FR4 (fiberglass insulator) being the most common. The relative dielectric constant of FR4 is $\varepsilon_r = 4.8$ (and $1/\sqrt{\varepsilon_r} = 0.46$). This means that the relative speed of electric waves propagating along a track on FR4 board will be 0.46 times the speed of light.

In particular, the PCB track layout is critical. At high frequencies, for example, capacitance between tracks can cause a lower resonance frequency in a tuned circuit. There will be coupling between adjacent tracks, so avoid placing low level feedback signal tracks next to tracks carrying high voltage switching signals. On a double-sided board avoid passing a current sense signal below the thermal pad of a surface-mount MOSFET because the thermal pad is connected to the drain and this carries high voltage switching signals. Watch out for tracks on the opposite side of the board, or inner layers in the case of a multilayer board. It is better if the inner layers are used for the ground and power supply planes.

Ground planes are important and not well understood. A current flowing through a track above a ground plane will induce an equal current, but in reverse direction, into the ground plane immediately below. If the ground plane is not continuous (maybe there are track routes cut across it), the induced ground plane current cannot follow the track above and will have to deviate. This will create a current loop and behave like an antenna, causing high levels of EMI to be emitted. Avoiding noncontinuous ground planes will avoid these problems.

Surface-mount circuits can have reliability problems due to thermal expansion of the circuit board. Components firmly attached to the tracks with solder can be stressed if they do not have the same thermal expansion as the board.

Through-hole construction is becoming less common, due to the lower cost of surface-mount circuit board assembly and the improved performance due to less parasitic effects. This is leading to a reduced availability of through-hole components. For slow-speed circuit prototypes through-hole components are ideal for fault finding and fast construction.

11.3.1 Through-Hole PCBs

In a through-hole PCB, it is usual for an RF or high-speed digital circuit to have an earth plane on the component side of the board. In many cases, an LED driver can be considered as a high-speed digital circuit. The earth plane serves two purposes; it screens the components from tracks passing underneath, and it provides part of a low-loss transmission line. By using FR4 board in a standard thickness of 1.6 mm, 50-Ω transmission lines can be created by making the printed circuit tracks 2.5-mm wide. A transmission line is formed between the earth plane and the track.

An earth plane on high speed PCBs may cause problems when an inductor is placed on the board because of the capacitive coupling between the ends of the inductor and the earth plane. This capacitance forms a parallel tuned circuit with the inductance and may cause the filter to be detuned. One solution is to remove the earth plane from the area below the inductor. An alternative solution is to mount the inductor on spacers above the board, so reducing the capacitance.

11.3.2 Surface-Mount PCBs

Surface-mount components are used extensively in most circuits. Ceramic capacitors are common but can be damaged by mechanical stress caused by circuit board expansion due to temperature. One method of minimizing this problem is to use physically small devices: devices larger than 1812 (0.18 $\times$ 0.12 in.) should be avoided.

Ceramic capacitors should be protected with a moisture-resistant coating. If moisture is absorbed into the ceramic material, the capacitance value will change. Moisture can also be absorbed into plastic packages, so a conformal coating over the whole board is preferred.

Some consideration should be given to storage of components before boards are assembled; metalized sealed bags should be used, perhaps with desiccant material. This will prevent moisture being trapped into an assembled board and avert the risk of damage during soldering (as the moisture boils off). Each component is given a moisture sensitivity level (MSL) by the manufacturer. This gives board assemblers an idea how to use the components; if MSL = 1, no real care needs to be taken. However, if MSL = 3 (say for a large integrated circuit), the IC has to be put in an oven to remove moisture before assembly. If this is not done the IC package could be damaged by any moisture inside turning to vapor during the heating process in PCB assembly.

Via holes are used to connect tracks on opposite sides of the board or to act as heat conduits below thermal pads. Through-hole PCBs have plated through holes that are 1 mm or larger in diameter. Surface-mount boards do not need holes large enough for component leads; hence they tend to be smaller in diameter. Metalized "via" holes 0.3 mm in diameter are common (used to connect two tracks rather than for component leads).

A problem with via holes arises when the board is heated. Glass and epoxy board, for example, FR4 type, has a high coefficient of expansion at temperatures above 125C. Above 125C the board goes through its glass transition temperature and its coefficient of expansion is greater than normal; Z-axis expansion increases the thickness of the board, and can cause fractures between the tracks and the via hole pads.

Soldering causes a problem due to the heat applied to the board; in wave soldering the board is heated to about 300C, which is way above the glass transition temperature. To reduce the problem of "via-hole" damage, all plated through holes should have a wall thickness of 35 μm or more. Temperature cycling of completed boards also causes problems.

On the surface, there is a temperature coefficient mismatch between components and the board. Leadless chip carrier (LCC) devices have an expansion coefficient of 6 ppm/C, but for the board it is 14 ppm/C (below the glass transition temperature) in the *XY* plane. Above the glass transition temperature the PCB has a coefficient of expansion of 50 ppm/C. Again, temperature cycling strains the solder joints and can lead to failure. A small gull-wing IC does not have a problem in this respect because the leads can flex a little.

PCBs built on aluminum sheet are often used with power LEDs. They can also be used for higher power LED driver circuits. Traditionally, copper-clad invar has been used within some PCBs to restrain expansion and to distribute heat. This should be used with polyamide boards, rather than glass and epoxy types.

A lacquer over the PCB called "solder resist" is commonly used to restrain solder within a solder pad area. However, surface-mount ICs use smaller packages than conventional leaded devices, so thin tracks of solder resist placed between the pads are not practical.

PCBs that have a fine track pitch tend to have 0.05 μm gold plating. If the gold is thicker it causes embrittlement. Gold or nickel plating gives a flat surface and makes surface-mount component placing easier.

11.4 Operational Amplifiers and Comparators

The operational amplifier has DC characteristics that may change with temperature, but those most affected are the dc offset, bias current, etc. The AC characteristics are less affected by temperature.

The greatest problem is that the op-amp is not ideal. The ideal op-amp has infinite input impedance, zero output impedance, and a flat frequency response with linear phase. Most practical op-amps have very high input impedance and this does not usually cause problems. The output impedance is not zero, and can be up to about 100 ohms. This is not often a problem because negative feedback is used to limit the gain of the op-amp and this also makes the effective output impedance close to zero. There is however an assumption: that the gain-bandwidth of the op-amp is far higher than that required by the circuit. The output impedance rises if the circuit requires a gain-bandwidth that is close to the IC's limit.

If the op-amp has insufficient gain-bandwidth product, excessive phase shifts occur and the circuit can show peaking in the frequency response. Gains of 20 dB close to the cut-off frequency can occur unless care is taken in the design. A good frequency response can be obtained by utilizing an op-amp that has a gain-bandwidth product many times that of the circuit's bandwidth. A rule-of-thumb value is 10–100 times the bandwidth.

Comparators are used in many LED drivers, to detect the current level in a sense resistor. Comparators can be described as an op-amp with a digital output; they compare two voltages on their input and set the output high or low depending on whether the noninverting input is higher or lower than the inverting input. Often a comparator has some in-built hysteresis to prevent jitter when the two inputs are at or near the same potential.

One weakness of a comparator is that they invariably have some input offset voltage; this results in an error in the switching and limits the minimum voltage that can be used for a reference. For example, the current sense comparator in the HV9910 LED driver has an offset of about 30 mV, and the maximum threshold for switching is 250 mV, so the threshold range is 30–250 mV, and could give a dimming range of just over 8:1.

It is possible to build a comparator by using an op-amp with positive feedback. However, the output stage has been designed as a linear circuit and the slew rate is slower than a comparator's output.

11.5 High-Side Current Sense

Sometimes it is necessary to measure the current flowing in a circuit that is at high potential relative to ground (0 V). This calls for a special component called a high-side current sensor. This device measures the current flowing in the high voltage circuit and produces an output at low voltage, to interface with low voltage control circuits.

An example high-side current sense IC in the Microchip HV7800 is shown in Fig. 11.8. This is exceptional since it is capable of 450 V working. Lower voltage rating parts are available from other suppliers, notably Linear Technology's LTC6102, which is rated at 60 V (there is an HV-rated version capable of 100 V). Also, Maxim's MAX4070 bidirectional high-side current sense amplifier, which is capable of measuring current in power rails having a potential of up to 24 V.

The circuit in Fig. 11.8 uses a shunt resistor R_sense to drop a small voltage. In this case R_sense = 0R1, so 2 A flowing through it will produce 200 mV drop. The "LOAD" input is high impedance, so the 47K resistor R_protect will not drop any significant voltage and the voltage between V_{in} and LOAD will be 200 mV. Within the IC there is a level translating circuit that measures the input voltage difference and then outputs the same voltage (relative to ground or 0 V). So the voltage at the OUT terminal will be 200 mV, which can be used as a feedback signal in the LED current control IC.

The reason for resistor R_protect is to limit current flow if the voltage across R_sense is too high. There is an internal Zener diode ~5 V between V_{in} and LOAD, so if the output were shorted to ground a high voltage could be developed across R_sense, forcing high current through the Zener diode and causing damage. The series 47K resistor limits the Zener diode current to a safe level.

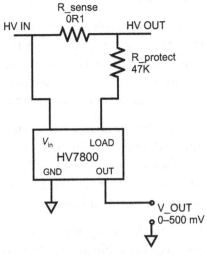

Figure 11.8: High-Side Current Sense.

CHAPTER 12

Magnetic Materials for Inductors and Transformers

Standard off-the-shelf transformers and inductors were described in Chapter 11. This chapter will describe magnetic materials and techniques for constructing custom transformers and inductors. If you have no interest in using anything other than standard off-the-shelf parts, this chapter can be skipped. The primary design requirement is to minimize losses, but to do this we have to consider copper losses, core losses, magnetic saturation, size, and construction. As this book is about designing light-emitting diode (LED) drivers, only the basics of magnetic materials will be given here. For more detail, the reader should consult specialist books on the subject.

An inductor can be made from a coil of wire, wound on a bobbin, and surrounded by a soft-magnetic core material. By soft, the meaning is that magnetization is easy and demagnetization occurs when the magnetizing force is removed. A hard-magnetic core is like a permanent magnet; it has high "remnance" (magnetic field remaining once the magnetizing force has been removed). Most magnetic materials have some remnance and it takes some reverse magnetic field to return the magnetic flux to zero; the field strength required to overcome any remnance, is called the "coercivity." On a graph showing magnetic flux versus field strength, the curve follows an italic "S" shape. But when the magnetic field direction is reversed, the flux does not follow the same curve; it needs more field strength (more energy) to return to the same point and thus forms a "fat S" shape. The fatter the S: the higher the magnetizing losses.

The core can be rectangular or cylindrical in cross-section with two halves that separate to allow the bobbin to be inserted. When the inductor is assembled, two spring steel clips (or adhesive) hold the two halves of the core together. This form of inductor is suitable for values of a few microhenries up to about 1 H. Some cores are toroidal (doughnut shaped) and are formed as a single piece, so that special coil winding machines are needed to wind a coil around the core.

The advantage of making a custom inductor is that they can be made to any value. Remembering that inductance is proportional to the number of turns squared, the number of turns required is given by the simple formula: $N = \sqrt{\dfrac{L(\text{nH})}{AL}}$. Here L is the required inductance in nanohenries and AL is the core's inductance factor (nanohenries per turn). Each core type has an AL value determined by the core manufacturer, which will be given in the manufacturer's datasheet or catalog. The AL factor is the inductance in nanohenries that will be produced for a single turn of wire.

The core's AL value is related to the permeability of the magnetic material used. Different magnetic materials are used, depending on the frequency at which the inductor is operating. If a particular AL value is required, it can be obtained by removing some of the magnetic material from the center of the core, thus creating an air gap. Note, an air gap in the center of the core, rather than in the outer material, reduces the emission of magnetic fields because the outer material behaves like a shield. The air gap has a lower permeability than the ferrite material, so increasing the gap reduces the overall AL value. A typical core gap is 0.1–0.5 mm, although it may be larger or smaller depending on the magnetic material permeability and the required AL value. The larger the air gap, the higher the magnetizing force that be achieved without saturating the core.

The presence of an air gap in inductor and transformer cores makes them suitable for high-magnetic saturation levels. An example application for this is inductors in power factor correction (PFC) circuits, which have a discontinuous magnetizing force. In PFC circuits, the current is switched on and off at high frequency with zero current flow between each pulse. The amplitude of the current pulse is made to rise and fall in proportion to the instantaneous AC voltage, so the average current is sinusoidal. Thus the power factor is close to unity (true sine wave).

Transformer cores that have no air gap are prone to saturate easily; their AL is normally far higher than an inductor core made from a similar magnetic material, but with an air gap. Gap-less inductor cores are often used in forward converters, in which the secondary current flows at the same time as the primary current. There is no stored energy in a transformer used in a forward converter.

If the coupling between windings must be very close, bifilar winding is often used. A bifilar winding has two insulated strands of wire twisted together before winding. Trifilar and higher order windings use multiple strands. However, if high-voltage insulation is required between the windings, bifilar techniques cannot be used unless special winding wire with high-voltage insulation is available, for example, Rubadue wire.

Sometimes multiple winding strands are used to reduce the equivalent series resistance because at high-switching frequencies the skin effect must be considered. Remember that the skin effect forces current to flow through the outside surface of a conductor, so if insulated strands are used the effective surface area is very large. A type of winding wire with multiple twisted strands is called Litz wire; each strand has a thin polymer film surrounding the conductor, for insulation.

12.1 Ferrite Cores

Ferrite cores are available in many shapes and material types. These cores are quite brittle and can break if dropped or struck with a hard object. Ferrite is usually a compound made from magnesium and zinc, or from nickel and zinc. Most ferrites have very poor-electrical conductivity, which limits any eddy currents in the core.

Nickel–zinc ferrites are used in inductors intended for EMI filters because they have high losses at high frequency; the core absorbs most of the energy above 20 MHz, up to about 1 GHz.

Manganese–zinc cores have losses that rise above 10 MHz, but have little effect on signals above 80 MHz. This characteristic makes them almost useless for EMI filtering.

Manufacturers data should be studied for details of the switching losses and optimum switching frequency. Ferrite is less effective at very low or very high frequencies. Generally, frequencies in the range 10 kHz–1 MHz are suitable for ferrite cores.

12.2 Iron Dust Cores

Iron dust cores (also called iron powder cores) are sometimes made toroidal shaped. The iron dust is ferrous oxide and is mixed in with clay-like slurry, which sets when baked. The result is ceramic material with soft-magnetic properties and with high-magnetic saturation levels.

These cores are good for switching frequencies up to about 400 kHz. From about 10 MHz, up to 20 MHz, the core is very lossy. Above 20 MHz the core has little effect and so cannot be used in EMI-filtering applications.

12.3 Special Cores

Proprietary compounds are used to make special cores. An example is molypermalloy powder (MPP). This has the ability to operate with high-flux density of typically 800 mT, rather than 200 mT of conventional ferrite cores.

MPP cores are distributed air gap toroidal cores made from a 79% nickel, 17% iron, and 4% molybdenum alloy powder for the lowest core losses of any powder core material.

MPP cores possess many outstanding magnetic characteristics, such as high-electrical resistance (thus, low-eddy current losses), low hysteresis (magnetizing) losses, excellent inductance stability after high-DC magnetization or under high-DC bias conditions, and minimal inductance shift when subject to flux densities up to 2000 G (200 mT) under AC conditions.

12.4 Core Shapes and Sizes

For custom inductors and transformers, E-cores are popular. An E-core has two halves that look like a capital E. The center segment is designed to pass through the middle of a bobbin on which the windings are wound. This center segment can be machined to create an air gap, as shown in Fig. 12.1, to allow high-magnetic flux without saturation of the core.

Variations on E-cores are EF and EFD cores. The EFD core is shaped so that the center segment is thinner than the main body of the core, so that the bobbin has a rectangular cross section, rather than square.

Pot-cores have a round body with a central spigot, so that a round bobbin drops inside the cavity. However, the area on the circuit board is essentially square. This means that the ferrite core has less material and does not provide the maximum use of the space. These cores are rarely used except in a tuned filter, when an adjuster is provided in the central spigot.

Toroidal cores are good from an EMI point of view because the magnetic field is fairly well kept in the ferrite core; there are no "corners" in the core where magnetic flux is prone to leak out. However, toroid shapes are difficult to wind, as the wire must loop many times through the central hole. Special coil winders are available for toroidal cores. Magnetic saturation can be a problem, so MPP and iron powder tend to be used because they have the ability to carry a high-flux density. A toroidal core is shown in Fig. 12.2.

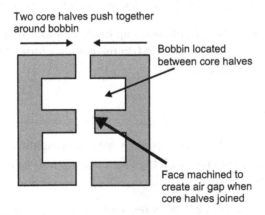

Figure 12.1: E-Core.

Figure 12.2: Toroidal Core.

12.5 Magnetic Saturation

Magnetizing (core) losses are also present and are due to the energy required to make the magnetic fields in the core to align with each other. In a switching circuit these losses are continuous and can cause core heating. These losses increase rapidly if the magnetization is forced to operate outside its linear region. Generally, the magnetic flux density should be limited to about 200 mT (200 Wb/m^2).

If an inductor or transformer has a large-discontinuous current flow, as in certain fly-back transformers and input inductors, the magnetic flux density may need to be lower than 200 mT. Ferrite core manufacturers recommend that flux variation due to ripple current or discontinuous mode operation should be limited to 50 mT. Inductors requiring the ability to handle high levels of flux variation sometimes use special cores with low losses at high-flux density, in which flux levels much greater than 50 mT are used. This allows a much smaller-inductor size.

The flux density is given by the equation: $B = \dfrac{L I}{N A_e}$. Here, L is the inductance, I is the peak current, N is the number of turns, and A_e is the effective core area. The inductance and peak current are calculated in the design of the LED driver circuit. We do not know the core area or the number of turns at this stage, but through iteration we can find something suitable.

The approach for choosing a suitable core is to select a core with a known effective area (A_e value), find the number of turns, and then calculate the maximum AL value that can be used with that size core. The number of turns can be found by transposing the previous equation: $N = \dfrac{L I}{B A_e}$. The equation for the maximum AL value is: $AL_{max} = \dfrac{L \cdot 10^9}{N^2}$.

Cores are usually available with standard AL sizes. If a core is available with a slightly lower-AL value than the maximum previously calculated, it should be selected and then a new value for the number of turns should be used, $N_1 = \dfrac{L I}{B A_e}$. However, if a lower-AL value is not available, a larger-core size with a higher-A_e value should be selected and the aforementioned process repeated. A simple spreadsheet can be created to make this process quick and simple.

Table 12.1: Wire resistance versus temperature

Temperature (°C)	Multiplying Factor
20	1.000
40	1.079
60	1.157
80	1.236
100	1.314

12.6 Copper Losses

Copper loss is the term used to describe the energy dissipated by resistance in the wire used to wind a coil. In 99.9% of cases this wire will be made of copper, whose resistivity at 20°C is about 1.73×10^{-8} Ω m. However, coils often have to operate above room temperature and will be heated by the operating losses in any case. The wire resistance at any temperature can be estimated from Table 12.1, developed by Mullard (now Philips).

Unfortunately, the resistance of wire also increases as the frequency of signals passing through it increases. The phenomenon of the "skin effect" is when the magnetic field caused by the current flow tends to force the electrons to flow down the outside of the wire. An alternating magnetic field produced by the current in the wire induces an electric field, strongest at the center of the wire, which repels the electrons and forces them to the outside surface of the wire. Thus changes in current produces a force that opposes those changes, which is inductance on a small scale.

The skin depth is given in Table 12.2.

Fortunately, Terman has created a formula for a wire gauge (in millimeters) where the skin effect increases resistance by 10%, which is a nominal limit that allows reasonable losses:

$$D = \frac{200}{\sqrt{f}} \text{ mm.}$$

For example, suppose we are operating at 100 kHz, then $D = 0.63$ mm. Using a larger-diameter wire than this does not give much benefit because the current will not be carried in the center

Table 12.2: Skin depth versus frequency

Frequency	Skin Depth (mm)
50 Hz	9.36
1 kHz	2.09
100 kHz	0.209
1 MHz	0.0662
10 MHz	0.0209

of the wire. In fact, in an LED driver (or any PWM power supply) harmonics are present at many times the switching frequency. In the earlier case, a significant proportion of the signal will have a frequency of 300 kHz.

In some cases, it is necessary to suffer higher-copper losses that are desirable, to have a transformer of a reasonable size. The use of Litz wire may be justified (although it is expensive) if low-copper loss is essential at high-switching frequency.

EMI and EMC Issues

Chapter Outline

The first two questions regarding electromagnetic interference (EMI) and electromagnetic compatibility (EMC) are: what is the difference between EMI and EMC? And which standards apply? Subsequent questions relate to how equipment can be made to meet the standards. Of course, meeting the standards often costs money (filter components, screening, and suppressors) so the aim is to just meet the standards with a small safety margin.

EMI is electromagnetic interference. This is the amount of radiation emitted by some equipment when it is operating. EMI is caused by emissions in the radio spectrum, which not only interfere with radio systems but also can cause other equipment to malfunction. One example is interference from portable radio transmitters like CB radios and cell phones; when used near a gasoline station, the pump can indicate the wrong amount being delivered. An often seen warning notice at a gasoline station says "using a radio transmitter can cause a fire," but in reality the most likely effect is to cause an error in the fuel measurement.

So what is EMC? This is electromagnetic compatibility, and is a measure of how good a system is at rejecting interference from others. Medical systems have a high immunity requirement because the consequences of a failure are death or injury. Any system connected to the AC mains power line must be immune to transient surges; the degree of immunity depends on the application. Power meters connected to lines where they enter a building are subject to the highest potential surges, so they have very high immunity requirements. Internal lighting and domestic appliances have very much lower immunity requirements.

Before we look at EMI and EMC standards, and the design techniques used to meet them, it is important that we understand signals. Fourier analysis shows that any signal that is not a pure sine wave can be considered as a fundamental signal plus higher frequency harmonics,

Power Supplies for LED Driving
249

Table 13.1: Harmonic Limits for EN 61000-6-3 Class C

Harmonic Order "N"	Maximum Current, Class C (% of Fundamental Current)
2	2
3	30 × power factor
4–40 (even)	Not specified
5	10
7	7
9	5
11–39 (odd)	3

which are a multiple of the fundamental frequency. For example, a square wave with a 50/50 duty cycle has a fundamental signal at the switching frequency plus a third harmonic of 1/3 amplitude, plus a fifth harmonic at 1/5 amplitude, plus a seventh harmonic at 1/7 amplitude, etc. If the signal is not 50/50 duty cycle, or if the switching edges have some slope (as all practical signals do), then there will be both odd and even harmonics present and the amplitude of harmonics will be less predictable. Typically, this is like the signal across a MOSFET switch in an LED driver circuit.

13.1 EMI Standards

13.1.1 AC Mains–Connected LED Drivers

Any LED driver connected to AC mains supply has to meet the limit specified in harmonic current emissions standard IEC/EN 61000-3-2. Within this standard there are several classes and the one related to lighting is Class C. The harmonic emission limits specified in IEC/EN 61000-3-2, Ed. 2: 2000, up to the 40th harmonic, are listed in Table 13.1.

Conducted emission limits in the 150 kHz–30 MHz frequency range are specified in the standard IEC/EN 61000-6-3.

13.1.2 General Requirements for All Equipment

All LED drivers have to meet the radiated emissions standards. The standard is IEC/EN 61000-6-3, which covers the frequency range 30 MHz–1 GHz. This standard uses limits previously set by CISPR22 in the USA and by the European Norm EN55022. The limits given in CISPR22 and EN55022 standards were intended for computers and communications-related equipment, but these have been adopted as generic limits for all electronic products, including LED lighting.

The emission levels to meet EN55022/CISPR22 Class B are 30 dBμV/m in the frequency range 30–200 MHz. From 200 MHz to 1 GHz the emission level increases to 37 dBμV/m. These are the signal levels measured at a range of 10 m from the equipment under test (EUT).

Since the signal power is proportional to $1/R^2$; for example, at 1 m from the EUT the emission limit will be 20 dB higher (100 times the power), at 50 and 57 dBμV/m, respectively.

13.2 Good EMI Design Techniques

It is important to look at the circuit diagram and determine where the possible sources of EMI are located. This should happen before the printed circuit board (PCB) is designed. The center point for EMI sources must be the MOSFET switch. This turns on very quickly and so has sharp edges with high frequency content. When looking at the circuit schematic, consider the effect of high frequencies (1–200 MHz).

At very high frequencies, components do not behave the same as they would at low frequencies. An inductor that was thought to block AC signals suddenly behaves like a capacitor that passes AC signals very easily. Similarly, a capacitor thought to have a low impedance characteristic behaves like an inductor at very high frequency; a good example of this is an electrolytic capacitor. So, check the component datasheets and look at the frequency response curves showing impedance versus frequency; see where the resonant frequency is—you will be surprised!

13.2.1 Buck Circuit Example

Let us take a look at a simple buck circuit, to see where the EMI can arise. Fig. 13.1 shows a typical buck circuit. As a reminder, the integrated circuit at the heart of this buck circuit is

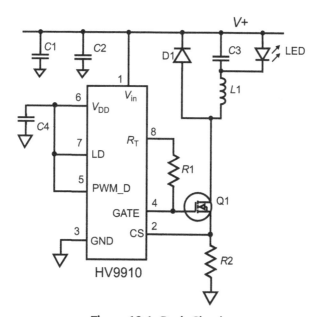

Figure 13.1: Buck Circuit.

a PWM controller. Internally, a clock signal triggers a latch, causing the gate drive output to be activated; the voltage on the Gate pin rises to 7.5 V in this case. The MOSFET Q1 turns on and the current increases at a fairly constant rate, due to the inductance of $L1$. When the voltage on the CS pin is raised above 250 mV, due to current in R2, the internal latch is reset and the gate drive output is disabled. The MOSFET Q1 turns off but current continues to flow in the LED and the flywheel diode D1 due to the energy stored in the inductor $L1$. When used in a buck circuit, this IC maintains an almost constant current in the LED.

For EMI analysis, we should consider what happens when the MOSFET Q1 turns on. The drain voltage of Q1 will fall to a very low voltage relative to ground, just a small voltage due to the current flowing in the drain–source channel of Q1 and in the current sense resistor $R2$. When Q1 turns off, the flywheel diode D1 conducts, so the drain voltage of Q1 rises very fast and is clamped to the positive supply rail. Thus, when switching, the MOSFET drain voltage will have a rectangular waveform that is almost equal in amplitude to the supply voltage. The fast rising and falling edges create EMI with a broad spectrum of harmonics.

Current flows are shown in Fig. 13.2. Analysis shows that the gate current flows from ground, through V_{DD} supply capacitor $C4$, through the IC and out of the gate drive pin, through the gate and current sense resistor and back to ground. Analysis of the LED current shows a path from ground, through the decoupling capacitors $C1$ and $C2$, through the LED and inductor, through Q1 and the current sense resistor $R2$, and back to ground. Both currents have fast rising and falling edges, which can produce high frequency EMI.

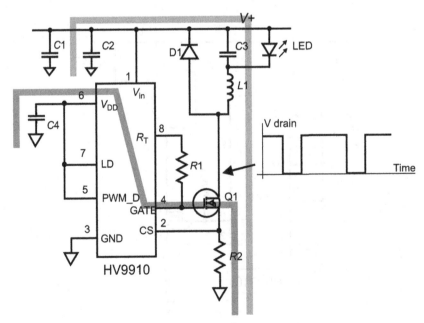

Figure 13.2: Buck Circuit Current Flows.

What is not shown in Fig. 13.2 is the current that flows through the flywheel diode D1. There is a forward current through D1 when Q1 is off, due to the energy stored in the inductor, which keeps the LED current flowing. There is also a momentary reverse current that flows in D1 when Q1 first turns on. This reverse current flows for a short time (typically 75 ns or less) and creates a current spike through the current sense resistor, R2. A small part of this current is due to the junction capacitance, but the main part is reverse recovery current.

Reverse recovery current occurs when a diode junction that is conducting in the forward direction is subject to a sudden reverse polarity. When a reverse polarity is applied, the free electrons in the junction area take some time to be swept away. Only after the free electrons have been removed is a depletion region created inside the silicon, which blocks further electron flow. The time taken to remove the free electrons in the reverse recovery time is T_{rr}.

The choice of capacitors in the circuit is important. Capacitor C2 must have low impedance at high frequency, for handling the high frequency current in the power switching circuit. The capacitor dielectric could be ceramic for low voltage supplies, or metalized plastic film, such as polyester in high voltage circuits.

The capacitor C3 mounted across the LED terminals provides a bypass path to carry high frequency signals, which are created when the MOSFET Q1 turns on charges the capacitance of the inductor windings. The inductors' winding capacitance is simply due to insulated wires being wound over each other in a coil. Some inductors have more self-capacitance than others, due to differences in construction. It must be of low impedance and be able to carry high frequency signals. A typical value is 100 nF. The voltage rating of capacitor C3 should be high enough to withstand the supply voltage because there is a chance that the LED could be disconnected or become open circuit.

The V_{DD} capacitor C4 should be a ceramic dielectric type, value typically 2.2 μF. This can be a low voltage type; a 16 V rating is commonly used. Remember that the actual effective capacitance with a 7.5 V V_{DD} supply will be lower than the nominal capacitance value, due to the bias voltage effect described in Chapter 11.

We briefly mentioned the inductor L1. We discussed the interwinding capacitance, which affects performance by causing current spikes when Q1 turns on. But the magnetic field must be considered too; a shielded inductor or a toroidal construction should be used to minimize radiating magnetic fields. A rod inductor makes a good antenna!

When considering input filters, we need to raise the impedance of the current path from the buck circuit into the power source by adding an inductor L2; this is shown in Fig. 13.3. A small capacitor C5 on the power source side of L2 shunts any small signals that manage to pass from the switching circuit and through L2. Basically adding L2 and C5 creates a low-pass filter to attenuate (reduce) and high frequency signals from the MOSFET Q1. Voltage spikes and noise on the supply rail are prevented from reaching the buck circuit by the attenuation due to L2, C1, and C2.

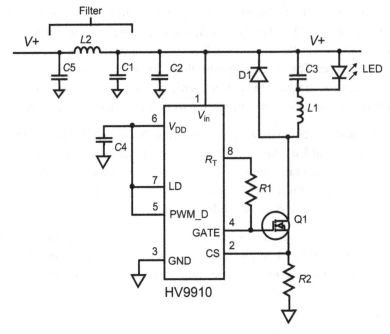

Figure 13.3: Buck Circuit With Filter.

If a filter has been added, but emissions are still too high, consider placing a resistor in series with the MOSFET gate. A value in the range 10–100 Ω is likely to be sufficient. The resistor slows down the MOSFET gate-charging rate when switching on and switching off, so the high voltage switching is slowed; the waveform now has sloped edges with fewer high frequency harmonics. However, slowing down the switching may also introduce more losses.

Using an oscilloscope to look at the voltage waveform across the current sense resistor in a high voltage buck circuit is revealing, see Fig. 13.4. First there is the turn-on spike,

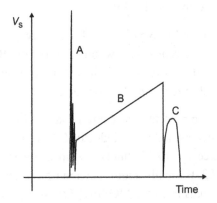

Figure 13.4: Current Sense Waveform.

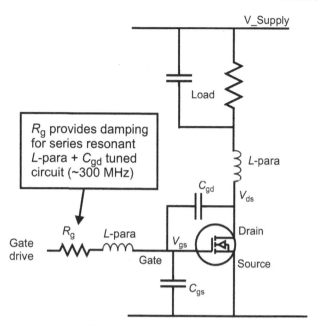

Figure 13.5: Gate Circuit.

"A," due to inductor parasitic capacitance and the flywheel diode reverse recovery period, but then there is immediately some oscillation afterward which is usually in the 100–300 MHz frequency range. After the oscillation, the current rises linearly, "B," due to the inductor. When the current sense threshold is reached, the MOSFET turns off, giving a sharp drop in the voltage across the current sense resistor. However, the MOSFET turns on again briefly, giving a final pulse of current, "C," through the sense resistor.

Some analysis of the circuit shows that the high frequency oscillation, after the initial turn-on pulse, is due to parasitic components resonating. Fig. 13.5 shows a circuit with the parasitic components added. Parasitic inductors are created by PCB tracks. A parasitic capacitor exists at the drain–gate junction of the MOSFET. The circuit is a series resonant LC circuit, which can be damped by a series gate resistor, R_g.

The short pulse of current, shown as "C" in Fig. 13.4, which occurs after the controller circuit has turned off the MOSFET, is caused by the MOSFET turning back on. Parasitic capacitance from drain to gate passes sufficient current when the MOSFET drain suddenly switches from almost 0 V up to a high voltage (may be 400 V). This current recharges the gate above the threshold voltage, allowing the MOSFET to turn on (at least partially).

The gate recharging effect is more prevalent in high voltage applications, where the MOSFET driver may not present a very low impedance at the gate. This effect will be made worst by a resistor in series with the gate because the impedance will be even higher. One solution is to add diode in parallel with the gate resistor, but this is only effective if the gate drive circuit has very

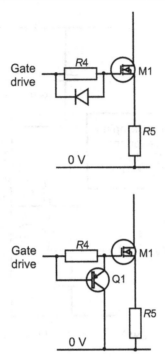

Figure 13.6: MOSFET Gate Drive Circuit.

low impedance and can clamp the gate to ground. An alternative solution is to add a PNP transistor across the gate resistor, with the emitter connected to the gate, the base connected to the gate driver, and the collector connected to ground (0 V). These two solutions are shown in Fig. 13.6.

When the gate driver output is being pulled to ground and current flows out of the MOSFET gate, the PNP transistor's base–emitter junction is then forward biased. The PNP transistor will turn on and the MOSFET gate will be pulled down to ground with low impedance; any current through the parasitic drain–gate capacitor will be shunted to ground. The collector–emitter voltage can drop to about 0.2 V at this time.

In some cases, the PNP transistor's base–emitter junction is reverse biased and could break down like a Zener diode when the gate driver output goes high. Some people add a signal diode in series with the base, to prevent reverse bias. However, many LED drivers have limited gate drive (the HV9910, e.g., only has 7.5 V drive and the current is limited to about 0.2 A), so these circuits will not have a problem. If an LED driver has a higher voltage/higher current gate drive output, it is unlikely to require the PNP transistor circuit.

PCB layout is also important to limit EMI radiation. The paths for the switching currents must be kept short and compact. If components cannot be colocated, so the path length is a little longer than desired, a return path should be placed alongside to ensure that the magnetic field from the current loop is minimized.

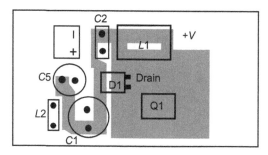

Figure 13.7: Printed Circuit Board (PCB) Bottom Layer.

It is important to realize that if a ground plane is beneath a conductor carrying high frequency signals, the ground plane will carry the return current along the same path. The reason for this effect is that magnetic forces produced by the signal current provide a lower impedance for the return current. These magnetic forces fall rapidly with distance from the conductor, so the lowest impedance path will be that closest to the signal path (i.e. directly underneath). It is therefore important that the ground plane is continuous along this path. If there is a break in the ground plane below the conductor, the current's return path may follow a circuitous route and then the ground plane will become an antenna!

Using the circuit schematic of Fig. 13.3, we can look at the PCB design. In Fig. 13.7, the track layout of the bottom layer is shown.

Notice how the ground connection goes from *C*5 to *C*1 and then on to *C*2 before reaching the ground plane. By avoiding the direct connection between the ground plane and *C*1, the current flow is steered in the direction we want. High frequency signals are taken from the grounded side of *C*2, which is low impedance at high frequency. The capacitors *C*5 and *C*1 hold up the input voltage during the cusps of the rectified AC input and are not intended to supply the high frequency current pulses needed for the LED load.

Fig. 13.8 shows the tracks on the PCB component side. The positive supply from the bridge rectifier BR1 flows to *C*5 and onto filter inductor *L*2. From the other side of *L*2, it passes to

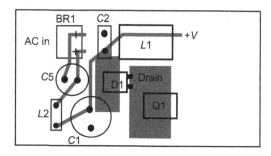

Figure 13.8: PCB Top Layer.

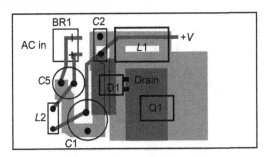

Figure 13.9: PCB Top and Bottom Layers.

*C*1 and then to *C*2. Notice that the *C*2 connection is a node where current also returns from the cathode of D1. Thus the high frequency flywheel current loop is from Q1 drain, through diode D1, and back to ground via *C*2; this is in a small area to keep the impedance low and EMI radiation to a minimum. As with the ground connection, the high frequency current path is kept away from the low frequency current flowing into capacitors *C*5 and *C*1.

Fig. 13.9 shows both sides of the circuit board overlaid. Notice that the earth plane is below the drain area of Q1 and inductor *L*1. Both Q1 and *L*1 have high frequency, high voltage switching, and a ground plane below helps to reduce the radiation from this area by screening underneath and making the node low impedance. Of course the parasitic capacitive coupling adds to the switching losses, but this cannot be avoided.

13.2.2 Ćuk Circuit Example

A Ćuk circuit is a boost–buck converter that performs well in a DC input application. An example of a Ćuk circuit is given in Fig. 13.10.

As with the buck circuit already described, or any other switching circuit, the aim in PCB design is to keep the switching currents flowing in as small a loop as possible. An earth plane under the main switching elements will also help reduce radiation.

Radiation couples easily into free space when the impedance of the signal source is similar (the impedance of free space is 377 Ω). Dipole antennas radiate and receive signals easily because their metallic elements are resonant at the transmit frequency and thus high impedance at the ends of the elements. Similarly, if the circuit area containing the high voltage switching signals is high impedance, it will radiate interference. An earth plane under the circuit lowers the impedance and reduces radiation. The PCB designer should take this into account when designing the circuit board.

High frequency emissions are caused by the fast rising and falling edges of the MOSFET drain voltage. As we discussed earlier, these can be reduced in amplitude by slowing down the switching of the MOSFET. Not only does this reduce high frequency emissions, it also

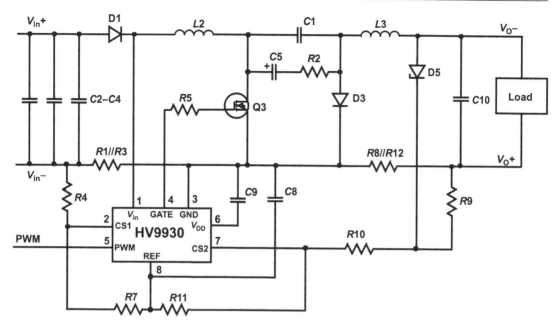

Figure 13.10: Basic Ćuk Circuit.

reduces high frequency ringing that is caused by the drain–gate capacitance resonating with stray circuit inductance. In Fig. 13.10 notice that a resistor ($R5$) has been connected in series with the gate. Slowing the MOSFET switching speed reduces the efficiency of the LED driver circuit, but saves the cost of additional filters.

Despite the natural input and output filtering inherent in the Ćuk circuit, additional filtering of the input and output power connections is likely to be required. A modified circuit is shown in Fig. 13.11, with an added input filter needed to meet demanding automotive specifications.

Capacitors $C3$ and $C4$ provide the current source for high frequency switching; these are ceramic capacitors and have very little high frequency ripple across them. Inductor $L1$ and capacitor $C2$ form a low pass filter to attenuate any ripple that appears across $C3$ and $C4$.

The circuit ground is not the same as the supply ground because two parallel resistors $R1$ and $R3$ break the ground connection. This means that a path for any high frequency signals is needed from the positive input to both the circuit ground and the supply ground. The path to circuit ground is provided by $C20$. This small value ceramic capacitor does not affect current sensing at the switching frequency. The path to supply ground is provided by $C2$, which is a high value ceramic capacitor.

Stray coupling from the LED driver circuit to ground can create a common mode signal that is present equally on positive and negative inputs. This means that a differential capacitor, like

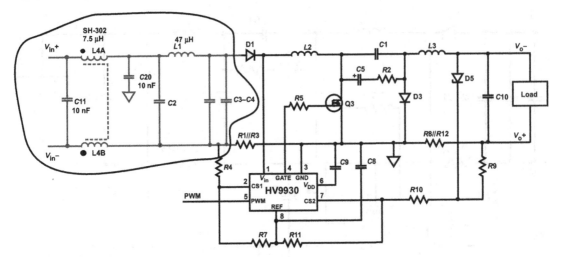

Figure 13.11: Input Filter.

C2, has no effect since the voltage is the same on both sides of the capacitor and no current flows through it. For this type of signal, a common mode inductor (choke) is required.

A common mode inductor *L4* has two windings on a common magnetic core. Differential currents produce opposing magnetic fields, so the result is no net inductance. Common mode currents produce magnetic fields that add together and thus have a high inductance. A common mode signal will present high impedance to common mode signals and thus reduce radiation.

Finally, a small value ceramic capacitor *C*11 is connected differentially across the power supply input to provide a low impedance path at the higher frequencies.

The output filter is highlighted in Fig. 13.12.

The output filter may be required, especially if there is a considerable wire length between the driver circuit and the LED load. If the distance is very short, the only filter usually required is a differential capacitor (*C*10) across the load. Distances greater than 10 cm (4 in.) between driver circuit and LED load can cause common mode signals to be created, primarily due to stray coupling between the LED and ground. Thus we may require a common mode inductor, *L*5, and a second differential capacitor *C*23. Small value ceramic capacitors *C*21 and *C*22 provide a shunt path to circuit ground for high frequency signals developed across *L*3 and the parallel current sense resistors *R*8 and *R*12.

In addition to a ground plane on the circuit board to reduce the impedance of the switching circuits at high frequency, a screen over the components may be needed. The position of such a screen is shown in Fig. 13.13.

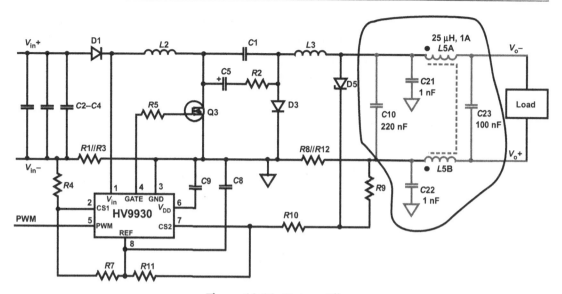

Figure 13.12: Output Filter.

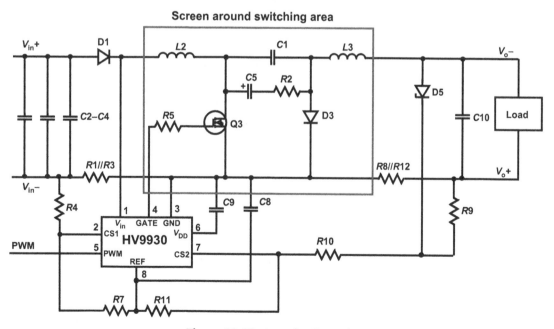

Figure 13.13: Area for Screening.

A metal screen over the switching area, with an earth plane underneath, provides an enclosure that stops EMI radiation. However, there will always be some leakage due to signals being carried outside the enclosure by connections to the remainder of the circuit. Even the PWM control wire will radiate unless a simple RC filter is added to it.

13.3 EMC Standards

The EMC performance is often automatically assured by the EMI precautions previously described. If radio frequency signals cannot get out of some equipment, they usually cannot get in either. However, electrostatic discharge (ESD) and surge immunity are two areas that are not taken into account in EMI practices.

People generate high electrostatic voltages during normal activities, such as walking across a carpet or opening a plastic envelope. A charged person touching electrical equipment can cause damage or malfunction. Thus equipment must be protected against high voltage discharge. Testing is carried out as specified in IEC/EN 61000-4-2 using an ESD gun, which can simulate the "human body model" (or HBM as it is often referred to). The standard voltage levels are 4 kV for a contact discharge and 8 kV for an air discharge.

Any equipment connected to the AC mains supply must withstand surge pulses, as specified in IEC/EN 61000-4-5. Each surge pulse has an open circuit rise time of 1.2 μs and a fall time of 50 μs. In domestic equipment, the peak surge voltage is 1 kV, which is added to the AC mains supply. In addition, 2 kV surges are applied between the inputs and ground (earth). The test pulses are positive and negative, and are applied at 0-, 90-, 180-, and 270-degree phases of the AC mains voltage.

Another form of surge test is the fast transient burst, as specified in IEC/EN 61000-4-4. This comprises ±2 kV pulses with a rise time of 5 ns and 50% decay at 50 ns. These pulses are repeated at a 5 kHz rate (200 μs between pulses), for 15 ms. There are 75 pulses in each burst, and the bursts are repeated every 300 ms, for 1 min. Testing is usually carried out by first applying ±250 V bursts, then ±500 V, then ±1 kV, and then finally ±2 kV.

13.4 EMC Practices

Equipment connected to AC mains power lines must be surge tested. The surges are applied, which are added to the normal AC voltage, at times to coincide with different phases of the AC line. The source impedance of the surge test pulse generator is a nominal 50 Ω. The energy in surge pulses can be absorbed or reflected to limit its damaging effects in the EUT. Absorbing the energy in surge pulses is the most common method of preventing damage.

A varistor, which is a voltage-dependent resistor made from a metal oxide, is commonly used to absorb energy by clamping the voltage. In a varistor rated at 275 V AC, the clamping voltage is typically 710 V, although conduction begins at about 430 V. The amount of energy absorbed in a varistor depends on its physical size. A varistor is usually wire-ended and disc-shaped; the diameter of the disc is related to the maximum energy (usually given in Joules). For example, a 9-mm disc varistor from Epcos that is rated for 275 V AC has a transient energy rating of 21 J and a peak current rating of 1200 A.

Another energy absorbing device is a transient voltage suppressor (TVS or TransZorb). This device is a Zener diode made in silicon and has a stronger clamping action. These are available with either bidirectional or unidirectional breakdown. In AC systems a bidirectional breakdown is required, but in automotive and other DC applications, a unidirectional breakdown is sufficient. TransZorb devices are usually rated in peak power (watts); 600- and 1500 W devices are commonly available.

The oldest technology, and still sometimes used, is the gas discharge tube. This has a glass tube filled with inert gas and metal electrodes at either end. When the voltage across the electrodes is high enough, the gas ionizes and conducts to clamp the voltage.

A plastic film capacitor is often connected across the AC line (typically 100 nF, 275 V AC X2 rated). This not only helps to reduce EMI emissions and susceptibility, it also helps to absorb some of the energy in surge pulses. Surge suppressors take some time to respond to impulse voltages; so fast transients can sometimes pass with little loss and can cause damage.

Many systems have a large electrolytic capacitor across the power rails, after a bridge rectifier. This capacitor will absorb surge energy; however, electrolytic capacitor construction results in some inductance that will have high reactance to fast rising surge pulses. A plastic film capacitor connected in parallel with the electrolytic capacitor will help to absorb high frequency energy. A clamping device, such as a varistor directly across the AC line is still a good idea because it limits the surge before it reaches the bridge rectifier.

A fuse should be fitted to every piece of equipment powered from the AC mains supply. This provides a means of limiting energy into an antisurge device, like a varistor. When a high-energy surge causes the varistor to break down, the fuse will blow. Some people fit a high power wire-wound resistor between the AC line and the varistor, to limit the current from surges and prevent burnout of the varistor.

When laying out a PCB, the spacing between tracks should be carefully considered. The breakdown voltage of an air gap is about 1 kV/mm, so at the potentially high voltage input of a power supply, sufficient gap should be allowed. An air gap of 3.2 mm is the minimum to prevent breakdown and a potential fire hazard. On a PCB the gap between conductors is known as the creepage distance. The air gap from a live part of the circuit to any other parts of the enclosure is known as the clearance distance.

Integrated circuits that can be powered directly from the rectified AC supply usually have "no connect" or NC pins adjacent to the high voltage pin. This is designed to give a suitable "creepage" distance. Where no gap exists, a slot can be cut in the PCB, or the contact pins can be coated with a conformal coating or a resin to increase the insulation.

CHAPTER 14

Thermal Considerations

Chapter Outline

14.1 Efficiency and Power Loss

People sometime refer to light-emitting diodes (LEDs) as being a cold-light source. This is true in the sense that an element is not heated to thousands of degrees Celsius to produce light. However, LEDs do indeed generate heat and this has been the cause of failure of several designs. As a first approximation, the heat generated is voltage drop multiplied by current flow. A white LED with a 3.5-V drop at 350 mA will produce about 1.225 W of heat. Actually the emission of photons (light) will reduce this power a little, but it is better to design a larger-heat sink to be on the safe side.

Power LEDs should always be mounted on a heat sink. For example, a traffic light using six or seven 1-W LEDs could be mounted alongside the driver electronics on a 6-in. (150 mm) diameter circular PCB. A heat sink could be mounted on the backside of the PCB for removing heat from both the LEDs and the driver electronics. As traffic lights may have to work in high-ambient temperatures, a good thermal conductivity is required.

When designing analog or switching power sources, we must always consider efficiency. This is the ratio power out/power in, and is usually expressed as a percentage. What designers sometimes overlook is that input power minus output power equals power loss in the LED driver circuit (Fig. 14.1). Loss in the driver must be dissipated as heat. A switching LED driver with 90% efficiency, driving a 10-W load will require an input power of: 10 W/0.8 = 11.1 W. This means that 1.1 W is the power loss and must be dissipated in the LED driver circuit, which raises the temperature of the circuit.

Another consideration is the maximum operating temperature of the LED driver circuit. If the ambient temperature around the circuit board is raised too high there could be reliability issues with components. Active components, such as integrated circuits, become less able to dissipate their internal power if the ambient temperature is high. Silicon-based ICs are usually okay up to 125°C ambient, although some are only rated to 85°C. Passive components also

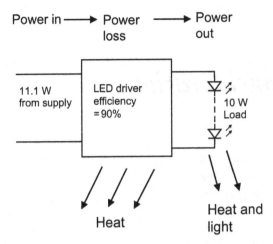

Figure 14.1: Power Loss in Driver Circuit.

need to dissipate heat, so a 0.25 W–rated resistor may have to be derated at high-ambient temperature. Also, as we discussed in Chapter 11, electrolytic capacitor lifetime will be shortened by high-ambient temperature.

14.2 Calculating Temperature

The temperature of a device can be calculated using simple "ohms law"–style mathematics. Temperature can be seen as being equivalent to a voltage. Thermal resistance can be equated to electrical resistance. Heat flow (watts) can be regarded as the equivalent to electrical current (Fig. 14.2).

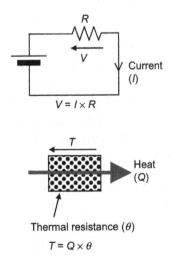

Figure 14.2: Electrical Equivalent Calculations.

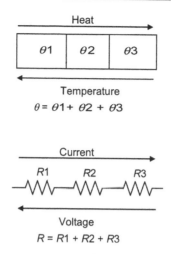

Figure 14.3: Thermal Resistances in Series.

Like electrical resistance, thermal resistances can be added when connected in series (Fig. 14.3). Consider a TO-220 package mounted onto an aluminum heat sink. The thermal resistance between the silicon die and the package, added to the thermal resistance of the package to heat sink interface and the thermal resistance of the heat sink to air interface, can all be added together to find the total thermal resistance from the silicon junction to air.

Thermal resistance is given as degrees Kelvin per watt of heat flow and has symbol theta. (Note, a 1K temperature rise = 1°C temperature rise.) The end points of the resistance are given as subscripts: junction (J), case (C), heat sink (H), and ambient (A). So, from junction to case, the thermal resistance is labeled as θ_{JC}. For example, let θ_{JC} = 1.2 K/W, θ_{CH} = 0.1 K/W, and θ_{HA} = 2.4 K/W, so the case to heat sink resistance is 0.1 K/W. The sum of thermal resistance from junction to ambient is 3.7 K/W; so when the device is dissipating 10 W, the silicon junction temperature will be 37 degrees hotter than the ambient temperature. If the ambient temperature is 25°C, the silicon junction temperature will be raised to 62°C.

Like electrical resistance, having parallel thermal resistance paths reduces the overall resistance (Fig. 14.4). Two paths, each of 2 K/W will create an effect single path of 1 K/W. This makes calculation of the exact temperature more difficult, as thermal paths are not usually as obvious as electrical paths. However, for a first approximation, calculating the temperature drop along obvious thermal paths will give a sufficiently accurate result. Less obvious thermal paths usually have a much higher-thermal resistance and have little effect on the temperature.

As parallel paths reduce the thermal resistance, in general a large-surface area can dissipate heat much better than a small area. Conversely, a small-surface area cannot dissipate a lot of heat. For this reason, a small driver is rarely able to drive a high-power load; remember this when the marketing department asks you to design a smaller driver!

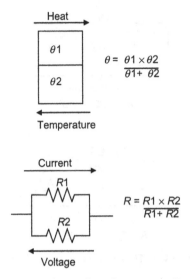

Figure 14.4: Thermal Resistances in Parallel.

Semiconductor component manufacturers usually specify the minimum and maximum junction operating temperatures for their devices. It is usual for the temperature range to be −40°C to +125°C, but this is not the ambient temperature. Commercial device ambient temperature ratings are 0–70°C, industrial device ratings are −40°C to +85°C. Military and automotive devices are rated for ambient temperatures of −55°C to +125°C, but these require special processing of the silicon material and packaging to achieve a +150°C junction operating temperature and hence are usually more expensive.

Component manufacturers also specify power dissipation (usually based on 25°C ambient temperature). Most manufacturers also provide thermal resistance information in their datasheets and some provide notes on heat sink requirements, which can be very useful to designers.

14.3 Handling Heat–Cooling Techniques

Heat must be dissipated somehow. If there is a high-thermal resistance from the source, the source temperature will rise until sufficient heat is dissipated (or until components are destroyed). High temperatures will reduce the reliability of components, so temperatures should be reduced somehow. One obvious cooling technique is to reduce the thermal resistance, and thus dissipate heat easier, by using a heat sink. This is fine if all the heat is generated in one place (like in a MOSFET or a voltage regulator).

Surface-mounted power MOSFETs are usually in a D-PAK or D2-PAK housing, which have a tab for dissipating heat. However, this means that the tab must be soldered to a copper surface area on the component side of the PCB, or otherwise a surface-mount heat sink is

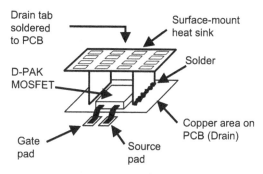

Figure 14.5: Surface-Mount Heat Sink.
PCB, Printed circuit board.

required (Fig. 14.5). A surface area of one square inch (25 × 25 mm) on a standard FR4 fiberglass board can give a thermal resistance of $\theta_{JA} = 30$ K/W with a D-PAK device. Surface-mount heat sinks are sometimes made from tinned brass and are soldered to the PCB, either side of the MOSFET body. These reduce the thermal resistance to about $\theta_{JA} = 15$ K/W.

Through-hole MOSFETs in a TO-220 package can be fitted to various heat sinks with a wide range of sizes. A small heat sink can be fitted, supported by the MOSFET pins or bolted onto the PCB through the TO-220 tab. Larger heat sinks could increase parasitic capacitance and cause an increase in switching losses, but this can be prevented if the heat sink is connected to the ground plane. Grounding the heat sink also prevents undue EMI radiation, but the MOSFET should be electrically isolated from the heat sink using a thermal conducting pad (made from a flexible material to give a large-surface contact). Switching losses will be due to the capacitance between the MOSFET drain (tab) and the heat sink.

Even where electrical isolation between the MOSFET and heat sink is not required, a thermally conductive pad or paste is a very good idea. This is because the surface of the MOSFET tab and the heat sink surface are not smooth. Without the thermally conductive pad or paste, the actual area where good contact can be made is just a small fraction of the surface available. Microcavities between the two surfaces creates air pockets with high-thermal resistance (Fig. 14.6). The thermally conductive pad or paste fills these cavities to create a uniform surface with low-thermal resistance.

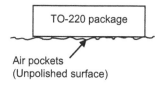

Heat sink

Figure 14.6: Thermal Resistance Created by Air Pockets.

Air flow bypasses inner fins

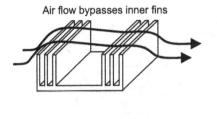

Air flows between the fins

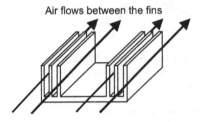

Figure 14.7: Fan Cooling of Heat Sinks.

When several devices on a circuit board generate the heat, a solution could be to use a fan to blow air across the circuit board. Cooler air from outside the equipment can be blown over warm components to reduce their temperature. Airflow will reduce the effective thermal resistance of the air interface.

Careful placing of cooling fans can make a big difference to the performance. Large objects, such as electrolytic capacitors, will tend to block the flow and may steer the cooling air away from areas of the PCB. If the air flows in the direction of heat sink fins, it will be more effective. Air flowing across the fins will only cool the front and rear fins (Fig. 14.7).

If mounting a fan at the top of equipment, make sure that the air flows upward, with the fan blowing air outward, so that the fan aids the natural buoyancy of the hot air. Fans mounted in the side of equipment are much more effective if two fans are used, one on either side of the enclosure. In a wide enclosure, both fans could be mounted on the rear panel; one fan should blow in and the other should blow out so that air circulates around the components inside.

Fans do have a reliability issue, so consider adding a fail-safe mechanism in case the fan fails to operate. A fail-safe mechanism should monitor the temperature of sensitive components on the circuit board. Driving the LEDs at a lower power or turning them off when the temperature rises too high may be a solution.

Safety Issues

Chapter Outline

This chapter discusses electrical safety and readers are advised to obtain the latest requirements from their regulatory body or safety consultant. The information here is to show that many topics must be considered, rather than as a reference for design work. Optical safety is a concern, but it is outside the scope of this book and readers are advised to consult technical data supplied by LED manufacturers.

15.1 AC Mains Isolation

Safety isolation can only be achieved with a transformer. This transformer can be placed on the AC mains supply, or as part of the switching regulator circuit. Transformer isolation on the AC mains supply is bulky because the AC signal is operating at 50 or 60 Hz.

Conversely, a transformer that isolates the *output* of a switching regulator can be very small because it is operating at the switching regulator frequency of typically 50 kHz or more. The use of multilayer insulated wire (also known as triple-insulated wire or Rubadue wire) is recommended for the transformer windings. When using triple-insulated wire, it is important to remember that such wire must satisfy the requirements described in Annex U of EN 60950. A simple example of an isolated LED driver circuit is given in Fig. 15.1.

If accurate current control is needed, additional electronics to control the LED current is needed and some form of isolated feedback is required. An optocoupler, which is an LED and a phototransistor in a single package, is usually used to provide the feedback signal. The LED in the optocoupler is powered from the transformer secondary winding and is controlled using a precision voltage reference and a current sense circuit, to accurately control the output

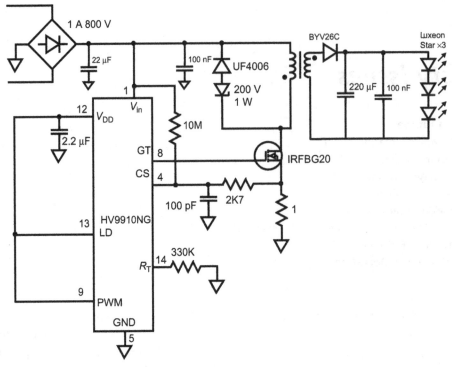

Figure 15.1: Isolated LED Driver.

current. The phototransistor enables and disables the switching of power to the primary winding, as required.

For products connected to AC mains supplies, 1500 V RMS (50 or 60 Hz) isolation is usually required. Products for medical applications usually require a higher isolation voltage with strict limits on leakage current; LED lamps are sometimes found in hospital operating theaters and in other medical applications.

15.2 Circuit Breakers

In the event of an overcurrent, the most common circuit breaker is a fuse. This is basically a piece of wire that is heated by the current flow. The wire eventually melts, thus breaking the circuit. In fast operating fuses, a weak spring is soldered to one end of the fuse wire; high current heats the fuse wire and as the heat softens the solder joint sufficiently, the spring wire pulls the fuse apart to prevent any arcing.

Fuses have a specified voltage rating. This is the maximum circuit voltage at which a fuse can be relied upon to safely interrupt an overcurrent. At voltages higher than the rating, a fuse may not be able to suppress the internal arcing that occurs after the fuse link melts. In electronic circuits, where limiting impedances ensure that the fault current is kept low so that a destructive arc cannot occur, fuses may sometimes be used beyond their specified voltage rating.

Fast-acting fuses are designed to open very quickly during an overload. They are not designed to withstand the inrush current or switch-on surge that occurs in some equipment. If an inrush limiter (NTC thermistor) is used, a fast-acting fuse gives the greatest protection. The overload current needed to cause a fast-acting fuse to open quickly (about 0.1 s) must be at least double the rated current. An overload of only 35% (e.g., 8 A through a 6-A fuse) could take an hour to open the fuse.

Time delay or slow blow fuses are designed to tolerate the temporary current surges that occur during switch-on, but they still blow quickly if subjected to severe overloads during faults. Slow blow fuses are commonly used in the primary circuits of electronic equipment, where the initial surge can be many times the normal load current. However, if a continuous overload occurs of double the rated current, the fuse will blow within 1 min.

Electronic circuit breakers are also available. These usually latch in the off state when a fault is detected, so that cycling the power supply off and then on again is generally required in order to reset the circuit breaker.

Tyco and others produce a type of fuse that becomes of high impedance when an overcurrent is detected, due to the current's heating effect, but then the electrical connection is restored once the fuse has cooled down.

15.3 Creepage Distance

In most electrical circuit connected to the AC mains supply, creepage distance is a concern. The concern is twofold: electrocution or fire. For example, a loose piece of solder could short out a pin carrying high voltage to another low-voltage point in the circuit; or moisture and dust could bridge the gap and allow a current to flow. In either example, the current may not be high enough to blow the fuse, but could be lethal to the user through electrocution or toxic smoke inhalation.

The requirements for creepage distance depend upon the application. Some integrated circuits have "no connect" (NC) pins between high and low voltage pins, so that a small piece of solder cannot bridge any two points. I have seen customers cut a slot in their circuit board to allow them to reduce the overall circuit board size. The creepage distance in air is much shorter than the creepage distance on a PCB surface. One way to avoid cutting slots in the PCB is to apply a conformal coating over the assembled PCB. The conformal coating is usually a silicone-based elastomer or a polyurethane varnish. Conformal coating materials must be UL recognized, particularly if a product is to be sold in North America.

When designing an isolated switching LED driver, a typical rule of thumb is to allow 8-mm creepage distance between primary and secondary circuits. A 4-mm creepage distance should be allowed between primary and ground. If these dimensions are allowed for during the design stage, there is a high probability that the product will gain approval with respect to creepage and clearance when the final product is submitted for test.

The working voltages within the circuit must be taken into account, not just the input and output voltages. Transistors and integrated circuits with built-in reinforced insulation (body thicker than 0.4 mm) must also still meet the spacing requirements at their connection pins.

15.4 Clearance Distance

Clearance is defined as the shortest distance between two conductive parts, or between a conductive part and the equipment enclosure, measured through air. Clearance distance helps prevent arcing between electrodes, caused by the ionization of air. The dielectric breakdown voltage depends on the relative humidity, temperature, and degree of pollution in the environment.

15.5 Working Voltages

The working voltage is the highest voltage that the insulation under consideration will be subject to when the equipment is operating at its rated voltage under normal use conditions. The appropriate creepage and clearance values can be determined from the figures provided in the relevant tables in EN 60950. These values must sometimes be calculated. To use Tables I–IV (2H, 2J, 2K, and 2L of the standard), the following factors must be considered: determination of working voltages, pollution degree of the environment, and the overvoltage category of the equipment's power source. The use of these tables is explained in Sections 2.10.3–2.10.4 of EN 60950.

When measuring working voltages, it is important to measure both peak and root mean square (RMS) voltages. The peak value is used to determine the clearance, and the RMS value is used to calculate creepage. Measurements should be accurate and repeatable and should also consider the end application.

15.5.1 Pollution Degrees and Overvoltages

Pollution degree is divided into four categories. The following definitions are based on those in IEC 60664:

1. *Pollution degree 1.* No pollution or only dry, nonconductive pollution occurs. The pollution has no affect on the circuit (e.g., sealed or potted products).

2. *Pollution degree 2*. Normally only nonconductive pollution occurs. Occasionally, a temporary conductivity caused by condensation must be expected (such as LED driver circuits used in a normal office environment).

3. *Pollution degree 3*. Conductive pollution occurs, or dry, nonconductive pollution occurs that becomes conductive due to expected condensation (such as an LED driver used in heavy industrial environment).

4. *Pollution degree 4*. Pollution generates persistent conductivity caused, for instance, by conductive dust or by rain or snow. Hopefully LED driver circuits will be protected from rain or snow!

Overvoltage is also divided into four categories according to IEC 60664:

1. *Overvoltage category I*. The lowest level hazard. Signal level voltages, usually in special equipment or parts of equipment.

2. *Overvoltage category II*. Local level voltages, such as in appliances and portable equipment.

3. *Overvoltage category III*. Fixed installations fed from the AC mains supply at distribution level. This could be something like a boiler fed from an internal building power feed.

4. *Overvoltage category IV*. Installations fed from a primary AC supply (overhead lines, cable systems, etc.). This category is usually applied to the power meter or main circuit breaker and is not relevant to most product standards.

Most standards are based on conditions being pollution degree 2 and overvoltage category II. Standard FR4 circuit board material is material group III. It is important to note that as working voltage, pollution degree, overvoltage category, and altitude increase, the creepage and clearance distances also increase. Table 15.1 gives the creepage distances for various working voltage for material group III because this is most commonly used for PCBs.

Table 15.1: Creepage distances

Working Voltage (Volts RMS or DC)	Pollution Degree 2 (Distance in mm)	Pollution Degree 3 (Distance in mm)
<50	1.2	1.9
100	1.4	2.2
125	1.5	2.4
150	1.6	2.5
200	2.0	2.5
250	2.5	4.0
300	3.2	5.0
400	4.0	6.3
600	5.3	10.0
800	8.0	12.5
1000	10.0	16.0

Linear interpolation should be used for working voltages between those listed. For example, for pollution degree 2, the distance is 2.5 mm for 250 V and 3.2 mm for 300 V. The increment is 0.7 mm over 50 V, so for 265 V operation the increase in creepage distance is 0.7 mm $\times$ 15/50 = 0.21 mm more than that for 250 V (thus 2.71 mm).

15.6 Capacitor Ratings

Capacitors connected across the AC line must be "X-rated," usually X2. These tend to be more expensive than standard capacitors because they are rated to withstand voltage surges. The typical DC operating voltage of X2 capacitors is 760 V, whereas the maximum DC level normally expected from a rectified 265 V AC supply is 375 V. Polyester or polypropylene (MKP) is the usual dielectric in X2 capacitors.

Capacitors connected from the AC line to earth must be "Y-rated," usually Y2. The typical DC operating voltage of Y2 capacitors is 1500 V. These capacitors normally have low capacitance (say 2.2 nF) and are usually made with a ceramic or polypropylene dielectric.

After the bridge rectifier, there is no requirement for a special voltage rating of the components because it is deemed that the bridge rectifier will become open circuit under fault conditions. In fact, the bond wires to the silicon die will act like a fuse.

Standard capacitors, rated at 400 or 450 V can be used on the DC side of the bridge rectifier. Since these are not rated for operation above their nominal working voltage, they are often smaller and of lower cost compared to X2 capacitors. For this reason some engineers will place the EMI filter after the bridge rectifier. However, an EMI filter before the bridge rectifier is preferred because it will tend to slow down voltage transients and prevent harmful voltage from reaching more sensitive components. It will also reduce the small transients generated by diodes in the bridge rectifier, caused by sudden changes in current flow during each half cycle of the AC supply, as capacitors on the DC side of the bridge rectifier are charged. A small X2-rated capacitor across the AC side of the bridge rectifier and the remainder of the EMI filter on the DC side of the bridge rectifier is a reasonable compromise.

15.7 Low Voltage Operation

The UL1310 Class 2 regulations and the European EN60950 safety standard (also known as IEC 60950) are generally applicable to any electronic circuit. The EN60950 standard was originally intended for Information Technology equipment, such as computers. However, since EN60950 is one of the few "harmonized" standards that have been agreed by all of Europe and many other countries in the world, it has been used as a reference for most safety regulations. If equipment complies with EN60950, it is deemed in law that due diligence has been performed.

The European Low Voltage Directive (LVD) is a safety regulation in Europe that covers all products operating from voltages of 50 to 100 V AC and 75 to 1500 V DC. There is a further "catch-all" General Product Safety Directive. These directives require a CE mark to be placed on all goods offered for sale. But to get permission to use the CE mark they must comply with safety standards like EN60950. Note that submodules do not require CE marking, but the overall equipment does. Clearly submodules ought to be safe to operate and any EMI should be low enough so that the final equipment can easily pass approval testing, otherwise the equipment assembler may decide to look elsewhere for his submodules!

To ease the burden in safety testing, many people ensure that their products operate at low voltage. The safety extra low voltage (SELV) requirements are that no touchable conducting parts have a voltage (relative to ground, or across any two points) above 60 V DC, or 42.4 V peak/30 V RMS AC. For example, a DC-powered (boost–buck) Ćuk converter, with 24 V DC input must not have an output above 36 V. This is because the Ćuk produces an inverted output, so the difference between the input and the output is the two voltages added together.

An AC mains–powered LED lamp must be transformer isolated to meet these regulations. The transformer must have primary and secondary windings with good galvanic isolation. As the primary circuit is an internal circuit connected directly to the external AC mains supply, this circuit contains hazardous voltages. The secondary circuit, which has no direct connection to primary power, may or may not be hazardous, depending on the secondary voltage.

Nonhazardous circuits are classified as SELV. The output voltage must be limited to below 60 V, which allows the electrical connections to be "touchable." If the equipment has an isolated cover, this is not enough to ignore the voltage in the primary circuit since the user could remove the cover. In the event of the cover being removed, there should be a microswitch to disable the equipment; alternatively, a second cover should be placed over the primary circuit. A double-fault (cover broken or removed AND microswitch broken or disabled) has to occur before the user can touch a potentially lethal voltage.

Each part of a circuit must be studied to determine the necessary insulation grade. Table 2G in EN 60950 describes common applications of insulation. By both measuring the working voltage and establishing the pollution degree, the appropriate row and column in Table 2H (and possibly also Table 2J) determine the minimum clearance distance required. For one test, the internal components and parts in both primary and secondary circuits are subjected to a steady force of 10 N, which may bend the PCB, but minimum clearance distances must still be maintained during the test. For example, to achieve the creepage and clearance demands between a primary circuit and a nongrounded SELV circuit usually requires reinforced insulation.

DC input products, however, can be treated in one of two ways. They can be considered as being fed by an extra-low-voltage circuit, or as hazardous secondary voltages. This would mean that the clearances could be calculated using Table III in EN 60950 rather than Table I, thus requiring slightly smaller clearance distances. DC input products may also be considered as being fed by SELV secondary circuits, depending upon the end application. If isolation is needed, then Table III of the standard is used. However, if isolation is not required in the end application, clearances can be ignored and only operational insulation is required.

As electronic circuits continue to get smaller, it is more important than ever to have a good PCB design that not only reduces electromagnetic interference emissions, but also reduces creepage and clearance problems. Where shortage of space is an issue, especially between primary and SELV circuits, techniques, such as slots cut through the PCB can be used to achieve the desired creepage distance. Slots must be wider than 1 mm to be considered acceptable. A lack of adequate creepage and clearance distance between a component in a primary circuit and a component in the SELV circuit is a common cause of product failure.

Control Systems

Chapter Outline

The original lighting control systems were analog: Triac phase dimmers and 1–10 V linear dimming. These have been obsoleted in many places by digital control systems using pulse width modulation (PWM) dimming. As the focus of this book is LED-driving power supplies, only a brief description of control systems can be given. Details of signaling protocols for digital systems, such as digital multiplex (DMX), digital addressable lighting interface (DALI), local interconnect network (LIN) bus, and CAN bus are complex and could fill a book on their own. In this book, I will give an outline of the uses and capabilities for the systems, with a little detail of the physical method of data transmission.

Analog control systems need to be extensively rewired if a change to the lighting control is required. So the advantages of using a digital bus controlled systems are: (1) the amount of wiring is greatly reduced and (2) the flexibility of being able to change the control system without rewiring. An example of a digital control system is office lighting controlled by DALI. There is a pair of wires (the bus) connected to all the lights in the office and a wall switch that can be used to turn off all the lights. But if the office is changed by the addition of a partition wall, the control system can be reprogrammed so that this wall switch will only turn off the lights that are located on the same side of the wall. A separate wall switch can then be programmed to control the lights behind the partition wall.

16.1 Triac Dimming

Triac control of LED drivers can be complex because of the limitations of the triac circuit. These limitations are the ability of the triac to keep latched in the "on" state and the effects of the electromagnetic interference (EMI) filters surrounding the triac. This is explored in Chapter 5 in the discussion about buck driver circuits (see Section 5.8).

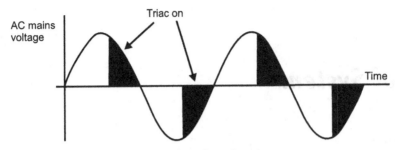

Figure 16.1: Triac Dimming.

In triac dimming, a control circuit in the light switch provides power to the lamps AC mains input for short periods of time each mains cycle (Fig. 16.1). Each half cycle has a period of 10 ms (for 50-Hz AC mains supply) and, in the example shown, the triac turns on after about 6 ms and off at the end of the half cycle. Thus the lamp is off at other times and so produces light with 100% modulation at a repetition frequency of 100 or 120 Hz. Old-style incandescent lamps produce light by heating a tungsten filament, so the 100% electrical modulation gives much less light modulation; this because the filament remains hot for some time and emits light until it cools down. However, unless an LED lamp has a large-storage capacitor that keeps the current flowing through the LED P–N junction, the LEDs turn off completely during the triac off periods.

The advantage of using a triac is that there is no additional wiring besides the AC mains supply, so triac dimming can be added to existing circuits with very little effort. The disadvantage is that the dimming range is limited and all lamps on the same circuit are dimmed together. It is still commonly found in residential lighting circuits.

16.2 1–10 V Dimming

The 1–10 V dimming system is analog and is sometimes referred to as 0–10 V dimming. The dimming range is actually 1–10 V, and 0 V turns the lamp off. This system is still in commercial use, dimming fluorescent tubes in offices and commercial buildings. However, it is gradually being replaced by DALI (see Section 16.3).

The 1–10 V dimming system has an advantage in that the AC mains can be permanently wired to all lamps in a building without any concern regarding the control. The control wires are connected separately, as shown in Fig. 16.2.

But there are disadvantages because it uses single-wire point-to-point control, with a common ground return. This means that if lamps need to be controlled separately, control cables with multiple wires must be installed, one for each lamp. If the control needs changing, such as if a partition wall has been installed in an office, the control system needs to be rewired. Single

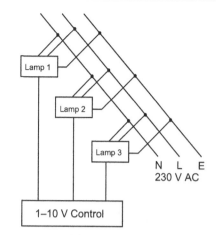

Figure 16.2: Wiring of 1–10 V System Lights.

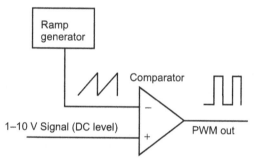

Figure 16.3: Converting 1–10 V Signals Into Pulse Width Modulation *(PWM)*.

wire control also means that it is susceptible to interference, so in electrically noisy industrial environments, the lamps could flicker when high-power equipment is used.

One method of interfacing a 1–10 V control system to an LED lamp driver is to use a ramp generator and a comparator, to create a PWM signal for controlling the light level. When the ramp voltage exceeds the 1–10 V signal, the comparator outputs a high-level output. This is shown in Fig. 16.3.

16.3 DALI

DALI is intended for general lighting, to replace the old 1–10 V linear dimming scheme. Installation is simple because a single pair of wires becomes a bus, with multiple light units connected to this bus. The control pair can be run from one lamp to the next, in parallel to the AC mains supply. As each light unit on the bus can have a separate address, a single pair of wires can be used to control 64 lamps independently. A typical wiring scheme is shown in Fig. 16.4.

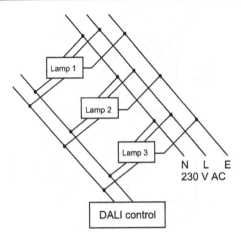

Figure 16.4: Wiring of Digital Addressable Lighting Interface (DALI) System Lights.

As the control bus is a balanced pair of wires and Manchester encoding is used for the signals, it is very reliable and not susceptible to external interference.

The DALI protocol uses logarithmic dimming, to match the sensitivity of the eye. Switching to half brightness means that the lamp actually appears half as bright to the eye and not that it produces half the amount of lumens. In fact, there are 256 steps of brightness, with the lowest level being 0% (off) and the highest level being 100% (full power). Due to the stepped levels, DALI is not suitable for theatre lighting or other lighting where the light level must increase smoothly in low-ambient light conditions. Theatre lighting needs a dimming range greater than 10,000:1.

16.4 DMX

DMX, sometimes known as DMX512, was developed for theatre lighting. The number 512 corresponds to the maximum number of data slots transmitted in each packet. DMX is also used in the entertainment industry in general for stage lighting during concerts and "disco"-style lighting in nightclubs. The reason for its use in these applications is speed; the 512 bytes of data can be transmitted in 23 ms but, by using shorter packets, the transmission time can be reduced to about 2 ms.

Data is transmitted over a twisted pair cable using EIA-485 (RS-485) differential signaling. For speed reasons, there is no error checking and correction of the transmitted data, but because it uses a very robust medium, which is not affected by external interference, errors are rare. Data lines must be terminated with 120 Ω, which is equal to the cable's characteristic impedance, to prevent signal reflections that would otherwise occur in an unterminated transmission line.

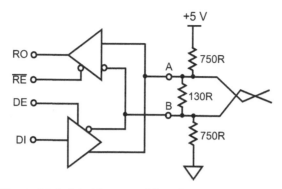

Figure 16.5: RS-485 Bus with Fail-Safe Termination.

RS-485 allows bidirectional half-duplex operation, with one driver connected at each end of the cable pair. Driver and receiver pairs can be located at various points along the bus, but only one driver can be active at any one time. The unused driver is put into the high-impedance state, so that it does not affect the data signals transmitted from the other end of the link.

The RS-485 bus should be one continuous pair with a 120-Ω terminating resistor at either end. An alternative "fail-safe" bus termination has a 130-Ω terminating resistor, a 750-Ω "pull-up" resistor from the A-wire to $+V_s$, and a 750-Ω "pull-down" resistor from the B-wire to ground. This ensures that the receiver is never in an indeterminate state; this fail-safe termination is shown in Fig. 16.5. Note that some RS-485 transceivers have a fail-safe mechanism built-in and therefore do not need the external bias circuit.

The RS-485 bus is a transmission line, so correct termination is important to prevent signal reflections. Spurs off the bus should not be allowed, unless the system operates at low speed. This is because reflections from an unterminated spur will affect the data pulse shape and cause errors. The RS-485 bus can be up to 1250-m long, at data rates of up to 100 kbps. At higher-data rates, the maximum line length is reduced. For example, at 10 Mbps, the maximum line length is about 30 m.

The RS-485 logic and voltage levels are +2 V to +6 V (Space) and —2 V to —6 V (Mark). Drivers are commonly powered from either ±3.3 V or ±5 V; for example, Intersil's ISL32458E can be powered from either of these voltages; it is very robust against 20-V common mode voltages and overvoltage conditions on the bus, with 60-V continuous voltage protection.

Like DALI, DMX also has 256 levels. However, two channels can be used together, so that one channel is used for coarse control and the other channel can be used for fine control. Thus the dimming range with two control channels is 65,536 levels.

16.5 LIN Bus

LIN is a low-cost serial bus for industrial and automotive applications. It is the second most popular in-vehicle network, after CAN Bus, with about one billion LIN nodes in use (in 2016) and growing.

The bus itself is single wire (LIN bus), with a ground return (GND) and a power rail (V_{Bat}). There is one master node and up to 15 slave nodes. The LIN bus voltage is in the range 9–16 V (the automotive battery voltage range). As the LIN bus is not a balanced transmission line, it has limited range and speed capability. The data rate is up to 20 kbps, transmitted over a distance of 40 m maximum. The master is identified by having a 1-kΩ termination on the LIN bus, whereas the slaves have a 30-kΩ termination. In practice, transceiver ICs have an internal 30-kΩ termination, so the slave nodes do not need any further termination on the LIN bus and only the master needs an addition 1-kΩ external termination. This is shown in Fig. 16.6.

There are many suppliers of LIN bus driver ICs. Examples are: ON-Semiconductor with NCV7382, Texas Instruments with TPIC1021A, and Microchip with MCP2021A. These all have a sleep mode, where the quiescent current is just a few microamperes, and slope control to limit the rise and fall time of the data pulses, to keep the EMI very low. The MCP2021A has the useful feature of an internal low drop-out regulator, to provide up to 70 mA for a microcontroller from the V_{Bat} supply.

Note that each supplier uses slightly different IC pin descriptions. Microchip use V_{BB} for the V_{Bat} supply pin, LBUS for the LIN bus pin, and V_{SS} for the ground (GND) pin. Texas Instruments use V_{sup} for the V_{Bat} supply pin, LIN for the LIN bus pin, and GND for the ground pin. ON-Semiconductor uses V_s for the V_{Bat} pin, BUS for the LIN bus pin, and GND for the

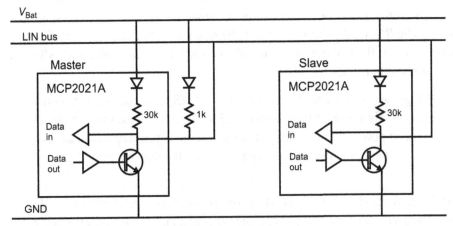

Figure 16.6: Local Interconnect Network *(LIN)* Bus Physical Connection.

ground pin. Fortunately, all three suppliers use the same RXD for the receive data pin and TXD for the transmit data pin. Other suppliers may have slightly different names.

As there is a common bus, with the voltage pulled low during signaling, it is possible to have a transmission clash. This is where two slaves attempt transmission at the same time. A clash is detected because a checksum is transmitted at the end of each message; if the checksum does not agree with the data received, the system (via the master node) asks each slave node to respond separately. Thus a clash causes several messages to be transmitted, with the consequential delay. As the LIN bus is not very fast anyway, it cannot be used for critical applications requiring high-speed control.

The most common use for the LIN bus is mechatronics, so in a vehicle this could include seat control, window control, etc. Another low-speed application could be turning on lights and adjusting their brightness.

16.6 CAN Bus

The CAN bus, like the LIN bus, is used in automotive and industrial control applications. The CAN bus is a much higher-speed serial bus, when compared to the LIN bus. One reason is that it uses differential signaling, which reduces transmitted EMI and reduces received noise, giving the system a high-level of electromagnetic compatibility. A variant of the CAN bus is "CAN high speed," which supports up to 1-Mbps transmission speeds, over cable lengths up to 40 m. By contrast, "CAN low speed" is another variant that supports data transmission speeds up to 125 kbps. CAN low speed has error checking and correction mechanisms, which account for the lower-data transmission speeds. The maximum transmission distance is 1 km, but at this length the maximum transmission speed is 50 kbps.

CAN bus drivers are powered from a 5-V supply. The maximum differential voltage across the bus is 3 V. If the differential voltage is greater than 1 V, it is said to be "dominant" and if it is less than 0.5 V it is said to be "recessive." The corresponding input and output data lines to the microcontroller are logic 0 (dominant) and logic 1 (recessive). The terms "dominant" and "recessive" will be described later. There are a number of manufacturers of CAN bus driver ICs, including Microchip (MCP2561), NXP (TJA1050), Maxim (MAX3058), Texas Instruments (SN65HVD232D), and many others. For those working on space applications, Intersil make radiation-hardened CAN bus drivers!

The CAN bus driver is connected to the bus, as shown in Fig. 16.7. Note, that the bus itself has 120-Ω line terminations, reducing reflections on the transmission line and thus helping to maintain speed.

Nodes are not allowed to transmit until there has been a period of inactivity on the bus. Once a quiet period has passed, every node has an equal opportunity to transmit. This is carrier

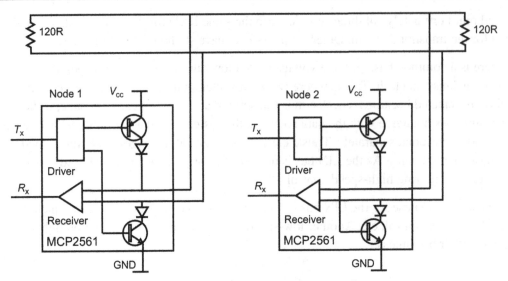

Figure 16.7: CAN Bus Physical Connection.

sense multiple access. When a node is transmitting a logic 1, it would expect to receive a logic 1, as the same pair of wires are used for send and receive. However, if it transmits a logic 1 and receives a logic 0, it knows that its transmission has been corrupted by another node transmitting at the same time. This is collision detection. Collision resolution is when the importance of the message is used to decide which node transmits first.

Earlier I mentioned the terms "dominant" and "recessive" are given in the CAN specifications. A recessive bit is a logic 1, and can be a level from an active or passive pull-up. The dominant bit is a logic 0 and is actively driven to ground by the transmitter. The idle state is the recessive level (logic 1). If one node is transmitting a dominant bit, while another node is transmitting a recessive bit then there is a collision. The dominant bit wins and the node transmitting the recessive bit identifies the collision and stops transmitting. Thus the higher-priority message is transmitted immediately and the node transmitting the lower-priority message automatically attempts to retransmit later, pausing for six clock cycles after the end of the dominant message.

Messages are addressed, not to the node, but to a function. A physical node carries data to a microcontroller, which processes the signals. A node and its microcontroller may support more than one function. Every node receives the messages and the microcontroller checks them for errors. If no errors are detected, the node will send an acknowledgment (ACK). Then the attached microcontroller decides if it supports the function being addressed. If no, the message is ignored, but if yes the message is acted upon. For example, one node could be located inside a car door may support the functions "unlock door" and "detect open door." If the car door has not closed properly, the "detect open door" function would signal that the door was "open," which would then be processed by the node controlling the display panel in the dashboard. The display panel node would light a warning lamp in the control panel.

The CAN bus is used to control LED lamps in automotive applications (and transport in general). Due to its speed, the CAN bus is used in some industrial applications. Although most industrial applications are mechatronic, some controlling lighting is also possible. For example, some elevators use a CAN bus for door control, motor control, and light control (including backlighting the push-button controls).

16.7 Wireless Control

Radio control systems can be Bluetooth, Wi-Fi at 2.4 or 5 GHz, or low-power radio using designated frequency bands, such as 433 MHz. The frequency of the system often determines the operating range because higher frequencies get absorbed more easily, so 433 MHz would normally have a much longer range than 2.4 GHz.

Regardless of the data transport mechanism, once a signal is received and decoded, it is converted into a physical medium like RS-485 for transmission to the lighting control system. The transmitted radio signal contains the lamp address and the light level setting required. Both DALI and DMX can be provided with radio interfaces. Examples of suppliers are Virtual Extension, which produces a DALI radio extender, and Wireless Solution Sweden AB, which produces a wireless DMX system.

Magnetek produces a wireless CAN bus transceiver, WIC-2402, which operates in the 2.4-GHz band. It has a 500-m range and a data speed up to 500 kilobaud, and so could be preferable to long-range wired connections.

Applications

Chapter Outline

The number of applications for LEDs is growing very fast, partly because other means of lighting are being phased out (often due to government legislation) and partly to save energy (money). Governments are trying to stop the use of lamps that contain hazardous materials or are very inefficient. Only a few specialist types of filament lamps are now produced and fluorescent tube lamps are becoming more difficult to find. People are still buying cold-cathode light bulb replacements, but they do take several seconds to reach full brightness, which some find annoying. Now LED lamps are preferred as their cost is quite low.

The use of LEDs is beneficial in applications where vibration and mechanical shock are present. A good example is lighting in automotive applications, such as motorcycle rear lamps or tractor work lights. In these applications, traditional filament lamps had a relatively short lifetime because of the vibration weakening the filament. Vibration does have some effect on the driver electronics; inductors have been known to break off the board after a mechanical shock, but they can be protected by conformal coating or potting in silicone resin. Using LED lamps has safety advantages because the light will almost always be available. The use of LED lamps also reduces maintenance time, thus increasing productivity.

The use of LED lighting in difficult to access places is beneficial, to reduce maintenance cost and inconvenience. Lamps that are mounted in public places, such as in the roof space of shopping centers and airports, require the area to be cordoned off and motorized hoists or platforms to be employed. These safety measures are inconvenient and expensive, so using LED lamps with a very long lifetime has long-term benefits that outweigh the initial costs.

17.1 Light Bulb Replacements

Light bulb replacements tend to be low power, but because they are much more efficient than incandescent filament lamps, they produce a similar light output. Here the design problems are small size, heat dissipation, and sometimes isolation. If the lamp has a metal heat sink that is electrically connected to the LEDs, isolation is essential. Lamps with a ceramic heat sink avoid this problem, allowing a lower-cost nonisolated driver to be used.

The vast majority of light bulb replacements are AC mains powered. Due to their low-power consumption, there is no legal requirement for power factor correction (PFC). The law requires lighting applications with 25 W or more power to have a good PFC, with a power factor of 0.95 or more. However, as described in Chapter 8, some local regulations and "green" organizations do have some more onerous PFC requirements (power factor = 0.7–0.9) for all LED lamps, regardless of their power consumption.

17.2 Tube Light Replacements

Tube lamps are made using long-glass tubes with a small diameter, typically 25 mm (1 in.). A metal cap at either end closes off the tube and allows electrical connections. The tube can be up to 2-m long, and this can make it difficult to achieve evenly spaced light distribution when using LEDs. So it is necessary to use either many LEDs, to create a continuous light, or use fewer LEDs with optics to diffuse the light evenly. As low-power LEDs are more efficient (10×100 mW LEDs produce more light than a single 1-W LED), having many LEDs not only makes continuous light along the tube length easier to achieve, but also increases the light output.

Tube lamps need the drive electronics to be housed inside the tube, but in such a way that light output is not obscured. Due to the tubular space, it is common to find a PCB strip across the center of the tube, with LEDs on one side and drive electronics on the other side. One potential problem is power dissipation because the circuit may draw 10 W power or more. A continuous glass tube may no longer be possible. The tube may have to be glass or transparent plastic on the LED side and metal or head-conducting opaque material on the other side, as shown in the cross-sectional diagram in Fig. 17.1.

17.3 Streetlights

Streetlights tend to be high power, when placed alongside a road, but sometimes lower-power lights are used in parks and pedestrian areas. Another form of streetlight is a low-power lamp that illuminates a street sign. Most streetlights are powered from high-voltage AC mains, but sometimes rechargeable batteries are used in solar-powered lamps. One example of a solar-powered streetlight is lighting at a bus stop, where generally there is a shelter

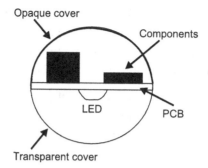

Figure 17.1: Cross-Section Light-Emitting Diode *(LED)* Tube Light.
PCB, Printed circuit board.

to protect against the weather with lighting inside; the solar panel can be mounted on the shelter roof.

Streetlights can be in the 25–250 W power range, with 100 W being common. In some cases the LEDs are required to be isolated from the AC mains, for safety reasons. For example, in Italy, it is a legal requirement that all streetlights are isolated. An isolated fly-back converter can be used in lower-power lamps up to 50 W, but fly-backs are not so efficient and a forward converter may be needed for 100 W or more.

As the power levels are needed by streetlights, they must also have good power factor. One nonisolated solution is to have a boost converter with constant on-time, producing ~400-V output, followed by a number of buck converters, each driving a long string of LEDs. By having several LED strings, there is some safety redundancy; if one fails, others will continue to produce light.

An example of where good PF may be required in low-power applications is street sign lighting. Street signs with "stop" or "give way" warnings often have a small lamp above them, so they can be easily seen at night. As the local authority has many of these lamps across a town or city, the combined power is considerably more than 25 W. Hence the local authority may demand a good PF for each lamp, so that the system as a whole has a good PF.

17.4 Theatre and Stage Lighting

Theatre and stage lighting is required to have a very high-light output, to illuminate the area effectively. On the other hand, because of the generally low-ambient light level, the dimming range has to be very wide. A range of 10,000:1 or more is typically required. This is to avoid the appearance of steps in the light level when lamp is initially off and then the light level is gradually increased.

The standard control protocol for stage and theatre lighting is DMX (see Chapter 16). The two main reasons for using DMX is these applications is the wide-dimming range with 65,536 levels and the fast response (23 ms maximum, 2 ms minimum). The problem for the LED driver is that such a wide-dimming range is very difficult to achieve.

Consider the dimming range of a simple buck circuit; the inductor is usually chosen so that the circuit operates in continuous conduction mode. This means that once the MOSFET switch turns off, current continues to flow through the inductor for perhaps four cycles before the stored energy is dissipated and current stops. The switching frequency of a buck controller is perhaps 1 MHz maximum, so a 1-μs switching period, which means that the smallest current pulse is about 4-μs long. To get 10,000:1 dimming, the longest pulse is 40-ms long. This means a PWM dimming frequency of 25 Hz would produce visible flicker. Generally, the PWM dimming frequency is 200 Hz (100 Hz absolute minimum).

One solution to give a high dimming range, if a buck circuit was the only option, would be to operate the buck convert continuously and use switches to bypass the LEDs when they need to be off. A short-circuit bypass would not be practical because this would imply zero duty cycle of the buck controller. There is a minimum switching time/minimum duty cycle to consider. However, it would be possible to have a "dummy load" whose voltage drop is less than the minimum forward voltage of the LEDs, but sufficiently high to meet the minimum on-time requirement. Such a scheme is shown in Fig. 17.2, where a normal P–N junction diodes acts as the dummy load. The disadvantage of this scheme is lack of efficiency, as it will draw moderate power when the LED is not lit.

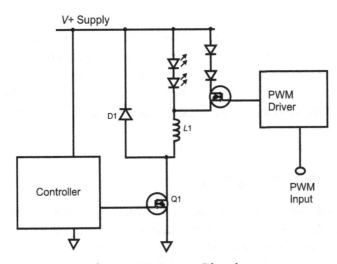

Figure 17.2: Bypass Dimming.
PWM, Pulse width modulation.

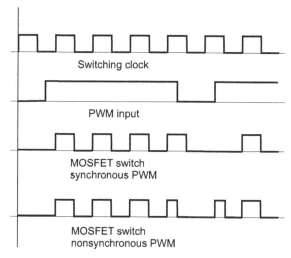

Figure 17.3: Synchronous PWM Dimming.

A boost circuit with an auxiliary MOSFET switch in series with the LEDs provides excellent dimming range. When the PWM dimming signal is low, the series MOSFET turns off and LED current stops instantaneously. To achieve good dimming, the boost circuit itself needs to have the switching clock synchronized with the PWM dimming signal, so that only whole switching cycles are possible. If the switching were asynchronous, it would be possible for fractions of a switching cycle to occur. For example, suppose that at the first-PWM dimming signal there were 2.5 switching pulses, but only 2.1 on the second-PWM dimming signal. As it takes a few cycles to fully charge the output capacitor, the current would change with each PWM pulse.

In Fig. 17.3, I show how fractions of a switching cycle can occur in nonsynchronous systems, if the PWM dimming signal rises or falls midcycle. In the synchronous system, the switching pulses will only be an output if the PWM dimming signal is high at the clock signal rising edge. Also, if the PWM dimming signal falls is the middle of a clock signal, the switching pulse remains high until the end of the switching cycle, so partial switching pulses cannot occur.

The ideal position would be for the output capacitor of the boost circuit to be kept fully charged by periodic switching, even when the PWM dimming signal remains off. This would prevent the LED current changing during partial switching cycles, caused by asynchronous PWM dimming. The difficulty in keeping the output capacitor charged is determining the output voltage; it should be just sufficient to achieve the correct output current when the PWM dimming signal goes high. In boost converter ICs, like the Microchip HV9912, there is an analog sample and hold function built in, which "remembers" the last output voltage. But this only holds up the reference voltage for a limited time. This is one case where a microcontroller wins over an analog system; the memory can be almost indefinite.

General stage lighting lamps do not need wide-dimming ranges. They can be either AC mains powered or battery powered, depending on the exact application. Those applications powered by AC mains may be nonisolated and drive lamps directly, or may be isolated using a built-in AC/DC converter and so effectively powered by a relatively low-voltage DC supply. Lighting a stairway, for example, may use a low-voltage DC supply because of the risk of being in contact with liquids.

17.5 Agriculture Lighting

In the past, plant growers used high-power ($\sim$500 W or more) HID flood lights to speed up plant growth. Now many use LED lamps for growing plants. This is because the LED lamp is far more efficient, allowing the running cost to be reduced. It has been found that only a few light colors affect plant growth, so part of the efficiency gain by using LED lamps is because the LED only outputs a narrow range of wavelengths and very little light is wasted. As only a few wavelengths affect plant growth, the white light from a flood lamp may have been appreciated by the gardeners, but not by the plants!

One advantage of LED lamps is that they run cooler than incandescent lamps and, as a result, plants need less watering! Even so, the LED lamps can be moderately high power, say up to 200 W, using many 1-, 3-, or 5-W LEDs in a large array. The highest power LED lamps for plant growing are about 1 kW, but lower-power lamps using more efficient LEDs are now available, so such power levels are very rare.

A disadvantage of using LED lamps with just blue and red light is that gardeners cannot tell if the plant is healthy. There is no green light to be reflected by the leaves; some growers insist of having a few green LEDs added, to give a little green light, for this reason. Blue LEDs with a wavelength around 460 nm are preferred for leaf and stem growth, but red LEDs with a wavelength around 610 nm are preferred for flower and fruit growth.

The demands on the LED driver are quite simple, as no dimming is required and there can be some tolerance on the LED current. The power levels required for the lamp will probably mean that AC mains power is necessary, with PFC (see Chapter 8). The humid atmosphere in a glasshouse will probably require the lamp to have AC mains isolation. A few specialist noncommercial applications, such as space travel, may use battery-powered LED lamps to grow plants.

17.6 Underwater Lighting

Lighting in and around swimming pools has to be powered from a low-voltage supply, for safety reasons. There are strict regulations for lights in swimming pools, see EN 60598-2-18.

Inside the pool itself, the regulations specify the use of 12 V DC maximum supplies, with low-leakage current and a power source 35-m away from the pool edge. Having an AC mains power supply so far from the pool edge means that a heavy-duty cable is required on the 12 V DC side, so that there is very little voltage drop along the cable. As 12 V DC is the maximum voltage at the pool, lamps are normally wired in parallel. Each lamp would then use a buck converter to drop the output voltage to a lower level, so typically the underwater lamp would contain only one or two LEDs.

Outside the pool, but within a few meters of it, safety electrical low-voltage circuits are a must. However, the voltage can be up to 50 V DC. Again, the power source must be 35 m from the pool edge. The AC mains supply to the power source must go via an 30-mA residual current detector circuit breaker, which disconnects the AC supply if there is an imbalance in the live and neutral current (i.e., an earth leakage).

17.7 Battery-Powered Lights

Battery-powered lights described here include those used in mobile phones, torches (flashlights) and other portable equipment. Vehicle lights and some theatre lights are also battery powered, but are described in Sections 17.4 and 17.9.

To get the most life out of a nonrechargeable battery, boost or boost–buck circuits are used. As an example, suppose that the output of a boost converter, driving the LEDs, is also used to provide power for the converter's PWM controller. The PWM controller would take power from the battery initially, but then take power from the higher-voltage output once the switching has started. The circuit would then be able to operate continuously until the battery voltage is almost zero.

A LED light that uses a single high-power white LED needs a forward voltage of about 3.3 V to operate, let us assume that the circuit is being powered by a battery of two alkaline cells (typically 1.5 V, but 1.6 V when new, and 1 V when spent). What are the driver topology options? A buck could not be used because the output voltage is slightly higher than the input. A boost could not be used because it needs to have some margin, because of duty cycle constraints, typically the output voltage needs to be 1.2 or more times the input. So a boost–buck would be a good choice in this case.

Cell phones usually use Li-ion batteries, where the cell voltage is about 3.6 V. Unlike alkaline batteries, the voltage across Li-ion types is only slightly lower until the energy is almost depleted. In other words, the terminal voltage during discharge does not change much. Note that rechargeable Li-ion batteries should not be operated with very low-terminal voltage because they can be damaged. Often an undervoltage lock-out circuit is used to disconnect the load, before the battery voltage falls low enough to cause damage.

In cell phones, the display (and sometimes the keyboard) is backlit. Due to the stable battery supply voltage, it is possible to use a white LED with just a series resistor as current limiter (Nokia used this technique). The brightness does not change very much as the battery charge reduces. However, for this scheme to work, the forward voltage drop of the LED is critical and LED manufacturers have to supply LEDs with a narrow-forward voltage drop. If the LED forward voltage drop is accurately controlled, LEDs can be connected in parallel.

Many cell phones now use boost or boost–buck topologies. This allows lower-cost LEDs to be wired in series, so the forward voltage matching is not required. They can be supplied from a number of different manufacturers, so long as the brightness levels are similar. Brightness is measured in lumens. The eye has a logarithmic amplitude response, so one LED can have almost double the lumen output of its neighbor before the difference is noticeable.

Sports and diving watches have backlights that often use a boost–buck circuit. The circuit has to be extremely small, but fortunately the power levels are very low and no large components are needed. The controller ICs include the MOSFET switches and are in small wafer-scale ball grid array or tiny DFN-type packages. The largest component is usually the inductor.

17.8 Signage and Channel Lighting

Channel lighting is defined as lights placed inside a (usually metal) space enclosed on three sides. These channels are formed to make letters for shop name boards, for example. The outward facing side is usually a colored plastic or glass cover. Before LEDs were used, neon tubes were bent to fit inside these channels. Now LED lamps can be made into almost any shape, which makes the job easier.

Channel lighting requires an isolated driver for safety reasons because of the metal channel housing. Most often, engineers use a standard AC/DC-isolated power supply with 12- or 24-V DC output. Then the LEDs are connected in a short string of three or six LEDs with a constant current driver in series. Some manufacturers produce such LED strings, with the LEDs and constant current circuit mounted on a short-PCB strip. The PCB strip can be either fiberglass board, such as FR4, or a flexible PCB that can be stuck directly onto the housing back panel.

17.9 Vehicle Lighting

Vehicles are not just cars and trucks. Lighting in trains, aircraft, caravans, forklift trucks, golf carts, tractors, cranes, and other industrial vehicles come under this category. However, usually only manufacturers of cars and trucks require lighting components to be AEC-Q100 qualified. To have a AEC-Q100 status, components are tested more rigorously. Semiconductors must have samples from three wafer lots tested, each for many months of temperature cycling. By comparison, standard semiconductors can be qualified with samples from one wafer lot tested for 1000 h. Consequently, automotive qualified parts can be more expensive. For that reason, manufacturers of caravans or forklift trucks would not necessarily want to use automotive-qualified parts.

Cars have traditionally used a 12-V lead acid battery as the main power source, but this has problems for design engineers. First, the 12 V is a nominal value and can be at 13.5 V during vehicle operation when the alternator is providing power. Voltage spikes can appear on the power rail too, due to inductive components being switched, but these are usually clamped to about 30 V by a transient suppressor fitted in the vehicle. During "cold-crank," when the vehicle is started on a cold morning, the battery voltage can drop to 6 V or lower. Due to the wide-battery operating range, boost–buck circuits are commonly used.

Some car-lighting systems use a boost converter, to raise the nominal 12 V up to a higher voltage, such as 36 V. This creates a lighting power bus. The system then uses multiple buck converters, one for each lamp, with the lighting power bus providing a power source. This technique allows low-cost buck converters to be used. Energy storage on the lighting power bus holds up the voltage during cold crank, so the lights do not flicker during starting.

Head-up displays use LEDs or laser diodes to project an image onto the inside of the windscreen. The image is then reflected back from the glass, for the driver to see. The images are to aid the driver, perhaps highlighting an area where children are playing near to the road, or to highlight a road junction ahead. At least one company was projecting the image from an infrared camera, so that people could be seen on a foggy day.

One LED/laser driver for head-up display systems is the ISL78365 from Intersil. It has four-constant current sink circuits; in a color system the four channels could be assigned to red, green, green, and blue (RGGB). The reason for the two green channels is that green LEDs are less efficient and so two are needed to attain the same light levels as red and blue. To make this driver efficient, the system is intended to have individual buck converters for each channel, to provide four positive supply rails. Each converter's output is controlled from the ISL78365, so that the headroom voltage on the current regulator is kept to a minimum and no energy is wasted. Fig. 17.4 shows this.

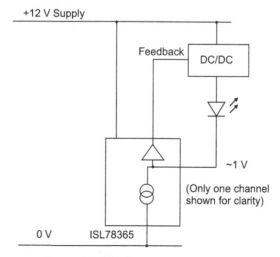

Figure 17.4: Head-Up Display Driver.

Recently, cars using a dual-battery system have started to appear. These use a 48-V battery and a 12-V battery. The two batteries can exchange power in either direction. These systems are likely to be found in cars that have stop–start motor control, where the engine is turned off when the car is stationary and the brake has been applied for a few seconds. The starter motor can be operated from the 48-V supply, to reduce the current drawn and hence voltage drops in the power cables. However, much of the car electronics needs 12 V and so the second battery provides this. But the lighting system could be powered from the 48-V battery, to allow simple low-cost buck converters to be used.

Trucks and buses can have 12- or 24-V batteries; 24 V is common in Europe. This means that electronic circuits used in a truck or bus has to withstand much higher-voltage transients. When the alternator is running and the battery is disconnected, a condition known as "load dump," the voltage can reach 120 V. The higher voltage helps in that a simple buck circuit is often sufficient, but it must be able to withstand much higher voltages.

Train voltages are variable. For example, metro trains are typically 110 V DC nominal (77–138 V). In Europe, many trains use a 72-V supply. Some diesel-powered trains have a 24-V supply. For the majority of lighting requirements, a high-voltage buck controller (like the HV9910 or its clones) could be used here.

Forklift trucks and golf carts have very large-lead acid batteries and voltages are typically in the 24–120 V range. The voltage can be much higher during charging, but as they do not have a starter motor, cold crank is not an issue. Of course, when motors are operated, such as lifting the "forks" in a forklift truck, there can be considerable transients. Again, a high-voltage buck controller would be a good choice here.

Aircrafts have an AC power supply, typically 110 V AC at 400 Hz. Cabin lighting and emergency exit signs using LEDs are now commonly used. Note that space-grade (radiation-hardened) components may be necessary because the aircraft flies at high altitude for most of its life. Cosmic rays/heavy ions can cause damage in integrated circuits, but thick metal enclosures can be used to give some protection.

Emergency exit strips in aircraft traditionally used along the side of the aisle are electroluminescent, with a phosphor material between two conducting layers (the top layer being translucent) and directly powered from the 400 Hz, 110 V AC supply. Some LED replacements have been used more recently and would typically use an AC/DC converter. However, photoluminescent strips that do not need any power source are also now being used.

17.10 Other Lighting

Emergency lighting systems often power the lights in exit signs, so if the power fails the exit sign can be clearly seen. Some AC mains office lights have an emergency light built in, for example, a $600 \times 600 \text{ mm}^2$ ceiling panel light may have an emergency light fitted in one corner; this allows the room occupants to see well enough to be safe. Such emergency lights are powered from a rechargeable battery that switches the LED lamp on when the AC supply fails.

In large systems, the rechargeable battery can be held centrally within the building, with a low-voltage bus connected to each emergency light. Smaller systems have a low-voltage bus with a current limiting circuit that prevent the bus voltage from being pulled down by any one of the emergency lights in the system. Thus the bus trickle charges the batteries continually and the batteries supply the power to the light for a short period only. A similar system is used in fire alarms, where the alarm sounder and the flashing beacon are powered from the battery, which has been trickle charged over the low-voltage bus.

Some alarm clocks have a light, instead of a buzzer or radio, to wake the user. Such alarm clocks have a special dimming requirement. As they are used in a darkened room, typically a bedroom, the dimming range has to be very wide. This is similar to the theatre lighting requirement and a dimming range of 10,000:1 or higher is required. When the light first turns on it must be very dim and gradually brighten to wake the user. A step effect in brightness is not acceptable; it must brighten smoothly like light from a rising sun. My comments in Section 17.4 about theatre lighting are also relevant here.

The "dawn" effect in an alarm clock is helped if the light is initially a yellow–white color or "warm white." Then, as the light gets brighter it should become a blue–white color or "cold white." This can be achieved by using warm-white LEDs and then adding blue light by switching blue LEDs into the circuit at high-brightness levels.

Bibliography

[1] M. Brown, Power Supply Cookbook, Newnes, Woburn, MA, 2001.

[2] A.I. Pressman, Switching Power Supply Design, McGraw-Hill, NY, 1998.

[3] K. Billings, Switch-Mode Power Supply Handbook, McGraw-Hill, NY, 1999.

[4] L.T. Harrison, Current Sources and Voltage References, Newnes, Burlington, MA, 2005.

[5] A. Zukauskas, M.S. Shur, R. Gaska, Introduction to Solid State Lighting, Wiley Inter-science, NY, 2002.

[6] G. Kervill, Practical Guide to the Low Voltage Directive, Newnes, Oxford, 1998.

[7] B. Rall, H. Zenkner, A. Gerfer, Trilogy of Inductors, Würth Elektronik/Swiridoff Verlag, Waldenburg Germany, 2006.

[8] Texas InstrumentsMagnetics Design Handbook, Texas Instruments Incorporated, Dallas, TX, 2001.

[9] M.I. Montrose, E.M. Nakauchi, Testing for EMC Compliance, Wiley Inter-science, NY, 2004.

[10] M.I. Montrose, Printed Circuit Board Design Techniques for EMC Compliance, Wiley Inter-science, NY, 2000.

[11] J.D. Lenk, Simplified Design of Switching Power Supplies, Butterworth Heinemann, Newton, MA, 1995.

[12] T. Williams, EMC for Product Designers: Meeting the European Directive, Newnes, Oxford, 2001.

[13] M. O'Hara, EMC at Component and PCB Level, Newnes, Oxford, 1998.

[14] A. Mednik, M. Tan, R. Tirumala, Supertex Application Notes: AN-H48, AN-H50, AN-H55 and AN-H58, Supertex Inc., Sunnyvale, CA, 2006 (now part of Microchip, www.microchip.com).

Index

Printed in the United States
By Bookmasters